A SHEARWATER BOOK

*The Forgotten Pollinators*

We may never know when early peoples first discovered that plants could be pollinated by hand without recourse to their natural pollinating agents. This carving, however, from the palace of Assyrian King Ashuir-nasir-pal II (83–859 B.C.) discovered at Calah, is the earliest known depiction of such an event. Male flowers held in the left hand are being used to dust and transfer pollen to the receptive female flowers of this date palm, one of the first domesticated crop plants. The original is now housed in the Metropolitan Museum of Art in New York City.

# The Forgotten Pollinators

*Stephen L. Buchmann*
*and Gary Paul Nabhan*

*Forgotten Pollinators Campaign,*
*Arizona-Sonora Desert Museum*

WITH A FOREWORD BY EDWARD O. WILSON
ILLUSTRATIONS BY PAUL MIROCHA

ISLAND PRESS / Shearwater Books
*Washington D.C. / Covelo, California*

Library of Congress Cataloging-in-Publication Data

Buchmann, Stephen L.
   The Forgotten pollinators / Stephen L. Buchmann and Gary Paul Nabhan ; illustrations by Paul Mirocha.
      p.   cm.
   Includes bibliographical references (p.      ) and index.
   ISBN 1-55963-352-2 (cloth).—ISBN 1-55963-353-0 (pbk)
   1. Pollination.   2. Animal-plant relationships.   3. Biological diversity.   I. Nabhan, Gary Paul.   II. Title.
QK926.B835   1996                                    96-802
574.5'24—dc20                                         CIP

*To the entomologists and botanists from Steve's years as a student: Phillip A. Adams, C. Eugene Jones, Jr., Robbin W. Thorp, Herbert G. Baker and Irene Baker, and E. Gorton Linsley. May your pollinator gardens flourish.*

*And to three scientist writers who have helped us remember the links between agricultural fields and the surrounding wildlands: Amadeo Rea, ethnozoologist; Efraim Hernandez Xolocotzi, agroecologist; and David Ehrenfeld, dean of conservation biologists.*

CONTENTS

I tracked down my father, who had wandered a little distance from the garden and was sitting against a tree trunk. In his fingers he carefully stretched out something that looked like a wasp, still alive. It was as broad as my hand, and had a yellow "8" on each clear wing, as plain as if some careful school child or God had painted it there. My father looked like he'd just had a look down Main Street, Heaven.

He told me, "There aren't any pollinators."

"What?"

"No insects here, to pollinate the garden. Look at this thing. How would it know what to do with a Kentucky Wonder bean?"

I couldn't know if he was right or wrong. I only faintly understood about pollination. I did know that the industrious bees did the most of it. "I guess we should have brought some bees over in our pockets too."

He looked at me like I was his spanking newborn baby; as if he loved me terribly but the world would never be what any of us had hoped for. "Rae Ann, honey," he said, "you can't bring the bees. You might as well bring the whole world over here with you, and there's no room for it."

"I know."

*Barbara Kingsolver*
TUCSON, ARIZONA

Great truths are sometimes so enveloping and exist in such plain view as to be invisible. One of them is the dominance on the land of flowering plants and insects. In addition to their overwhelming biomass, nearly a quarter million species of plants and three-quarters of a million insects have been described by biologists to date, together composing a full two-thirds of all kinds of organisms known to exist on the planet. Other terrestrial and freshwater groups, from protozoans to vertebrates, pale by comparison.

The joint hegemony of these two great groups is not an accident. It is the result of coevolution, the process in natural selection during which species adapt to one another and thereby build rich ecosystems. The interaction between plants and insects has been going on a long time. It began over 100 million years ago with the origin of the flowering plants and accelerated with their ascendancy in the world's vegetation during the early Cenozoic era some 40 million years later. Much of their coevolution was mutualistic: Species—more often whole complexes of species—came to seal obligatory partnerships with their insect counterparts. Such relationships, construed broadly, are among the central topics of ecology. Ants, for example, among the most abundant of insects, spread the seeds of plants, protect them from herbivores, and enrich the soil in which they grow. Insect detritovores such as termites and woodboring beetles convert dead vegetation into nutrients that can be reabsorbed by the roots of living plants. And, as Stephen Buchmann and Gary Nabhan so clearly and evocatively remind us, a majority of flowering plants must have insects to reproduce.

There is a welded chain of causal events that leads directly to our species: if plants, including many food and forage crops, as well as natural floras, must have insects to exist, then human beings must have insects to exist. And not just one or two kinds of insects, such as the friendly and lovable honeybees, but lots of insect species, vast numbers of them. The reason is that millions of years of coevolution have finely tuned the relations between particular plants and their special pollinators. The shapes and colors of the flowers, their scent, their location on the stalks, the season and daily schedule of their pollen and nectar offerings, as well as other qualities we admire but seldom understand, are adjusted precisely to attract particular species of insects; and those specialists in turn, whether beetles, butterflies, bees, or some other group, are genetically adapted to respond to certain kinds of flowers. In lesser numbers the same is true of the interactions between plants and species of birds, bats, and other vertebrates dependent on diets of pollen and nectar.

Nature, we learn, is kept productive and flexible by uncounted thousands of such partnerships. The connections are fragile, and we are reminded by melancholy case histories reviewed by Buchmann and Nabhan that when one partner is extinguished, the other is at the very least put at risk—and sometimes doomed, if it happens to be adapted to no other partner. No phenomenon in nature illustrates more vividly the principle that conservation measures must be directed at ecosystems, not just individual species. If the last pollinator species adapted to a plant is erased by pesticides, or habitat disturbance, the plant will soon follow. And as these and other populations decline or disappear, the consequences spread through the remainder of the food net, weakening other interspecific relationships.

Those unconcerned about the natural world, and I hope *their* numbers are dwindling by persuasion, will do well to consider the consequences for humanity of the decline of pollinator complexes. Eighty percent of the species of our food plants worldwide, we are informed, depend on pollination by animals, almost all of which are insects. One of every three mouthfuls of food we eat, and of the beverages we drink, are delivered to us roundabout by a volant bestiary of pollinators.

The evidence is overwhelming that wild pollinators are declining around the world. Most have already experienced a shrinking of range. Some have already suffered or face the imminent risk of total extinction. Their ranks are being thinned not just by habitat reduction and other familiar agents of impoverishment, but also by the disruption of the delicate "biofabric" of interactions that bind ecosystems together. Humanity, for its own sake, must attend to the forgotten pollinators and their countless dependent plant species. In the following chapters Buchmann and Nabhan make a compelling case for more focused research on pollinator complexes and increased attention to their status as an integral part of future conservation planning and restoration ecology.

*Edward O. Wilson*
MUSEUM OF COMPARATIVE ZOOLOGY
HARVARD UNIVERSITY

It has now been nearly five years since we brought together the first body of conservation scientists to address the dire consequences of having ignored plant/pollinator interactions in most discussions of biodiversity and agricultural stability. At a daylong symposium called "The Conservation of Mutualisms," we focused on the disrupted relationships between rare desert succulent plants and their pollinators, many of which have declined dramatically in abundance. The symposium's theme was considered important enough to attract cosponsorship from the IUCN Species Survival Commission, the International Organization for Succulent Plant Study, Bat Conservation International, and the Pew Scholars Program on Conservation and Environment. Hueblin, Inc., a distributor of products made from cultivated century plants—that is, tequila—helped underwrite travel for participants from three Latin American countries, out of its concern over the decline of wild century plants and their wild pollinators.

We are particularly grateful for the stimulus imparted by the participants in that first symposium, many of whom responded instantly to Steve's concept of "the forgotten pollinators." Many of them encouraged us to pursue this issue further: George Rabb, Ted Fleming, Vince Tepedino, Merlin Tuttle, Cathy Sahley, the late Maricela Sosa, Hector Arita, Abisai Garcia, Ted Anderson, Alberto Burquez, Luis Eguiarte, Robert Bye, Jr., among others. The editors of *Species* and *Conservation Biology* kindly printed highlights of that meeting in their journals, and these notices attracted additional responses from scientists around the world.

Since that time, we have had the benefit of financial and moral support from the Arizona-Sonora Desert Museum, the Wallace Genetic and Wallace Global Funds, the Geraldine R. Dodge Foundation, the Stocker Foundation, and Island Press. We are grateful for the collaboration of Charles Savitt, Barbara Dean, Barbara Youngblood, Lisa Magnino, Robert Wallace, Charlotte Fox, Scott McVay, and Victoria Shoemaker in seeing to it that this book is well integrated into a larger Forgotten Pollinators Campaign for enhanced public awareness and educational outreach. We thank our fellow writer/biologist and Tucson denizen, Barbara J. Kingsolver, for permission to quote her charming story about why you can't always bring the pollinators with you to newly established gardens.

Four people deserve special attention for blessing the campaign with their talents. Paul Mirocha's illustrations not only brighten the jacket illustration and pages within this book, but are a key element of the Forgotten Pollinators campaign's attempt to capture people's imaginations with the wonder and natural beauty and grace of pollinators worldwide. Mrill Ingram, our campaign coordinator, has given us valuable insights, advice, logistical support, and creative management skills since March 1995. David Hancocks, our museum's executive director, fully sanctioned the use of the museum's resources to fully explore the public interpretation of plant/animal interactions on the ground and through the media. In honoring us with a foreword, Edward O. Wilson, Pellegrino Professor of Science at Harvard University, has demonstrated that this topic should be an essential component of future discussions about global biodiversity and invertebrate conservation, causes he has so eloquently championed as a senior "ecostatesman" for decades.

We are also grateful to the campaign's board of advisers, informal consultants, and student interns: Robert Michael Pyle, Vince Tepedino, Peter Bernhardt, Judith Bronstein, Elizabeth Donnelly, Steve Walker, James Cane, Sergio Medellin, Phil Torchio, Beverly Rathke, Mark Dimmitt, Melody Allen, Merlin Tuttle, Andrew Matheson, Eva Crane, Nick Waser, Brien Meilleur, Peter Feinsinger, Gene Jones, Lucinda McDade, Bruce Pavlik, Steve Prchal, Michael Gregory, Carol Cochran, Cyndy Henzel, Rachel Levin, O. T. Kizer, Margaret McIntosh, Sarah Richardson. A special thanks to Cristina A. Bramley for tireless assistance with typing and innumerable last-minute format changes and editing as the manuscript went from our often non-cross-pollinating DOS and Macintosh computers to the reviewers and editorial staff at Island Press. We extend a special thanks to Don Yoder for improving our manuscript with his impeccable copyediting skills, and to Christine McGowan and Bill LaDue at Island Press.

Thanks also to others who have joined us in the field on diverse projects over the past years including Humberto Suzan, Marlo Buchmann, Marlyse Buchmann, Melissa Buchmann, Justin Schmidt, Hayward Spangler, James Hagler, the late Edward Southwick, Dini Eisikowitz, Robert Brooks, Doug Yanega, Bryan Danforth, the late George Eickwort, Robbin Thorp, Liz Slauson, Avi Shmida, Gloria Hoffmàn, Gene Jones, Marcus King, Karl Niklas, Charles Shipman, Jerome Rozen, Josh Tewksbury, Edward Wilson, Josh Kohn, John Tuxill, Mark Fishbein, Jono Miller, Mark Minno, Richard Felger, Robert Minckley, William Wcislo, John Alcock, Steve Thoenes, Karen Strickler, Bob Schmalzel, David Roubik, Luis Eguiarte, Caroline Wilson, James Cane, Charles Connor, Ellen Ordway, Veronique Delesalle, Christian Petrovich, Makhdzir Mardan, William Schaffer, Stephanie Zador, Lupe Malda, Evan Sugden, and Albert Jackman.

We especially want to thank the first cohort of Desert-Alert volunteers (coordinated by Cynthia Henzel) from the Arizona-Sonora Desert Museum. Should you be interested in volunteering for fieldwork, working alongside professional ecologist mentors, on rare desert plants and their pollinators, call the Arizona-Sonora Desert Museum in Tucson and ask for the Desert-Alert field course schedule. Stephen Buchmann's electronic mail (E-mail) address is buchmann@ccit.arizona.edu. Recently we added an electronic mail address for our Forgotten Pollinators public awareness campaign at the Arizona-Sonora Desert Museum. Our address there is: fpollen@azstarnet.com. By the time you read this, that address should be an active "listserv" on the Internet for those interested in pollination biology and our Forgotten Pollinators Campaign at the Desert Museum. We look forward to hearing from you and hope you will join our efforts through the Forgotten Pollinators Campaign—you may also wish to experience our pollinator gardens and examine related exhibits through a personal visit to the Arizona-Sonora Desert Museum a few miles west of Tucson, Arizona.

The best times to see the wildflowers along with the birds and the bees are in March/April and again in July/September following our summer monsoons when the desert truly smells like rain.

A special thanks to Mahkdzir Mardan for introducing us to the world of giant honeybees in peninsular Malaysia. We are indebted to Salleh Mohd Noor ("Pak Teh") and his honey hunting family for sharing his traditional honey gathering experiences with us, along with his passion for preserving these lowland rainforests and their Tualang trees.

Finally, we wish to acknowledge the many botanists, zoologists, farmers, beekeepers, honey hunters, restoration ecologists, and naturalists who have never forgotten the importance of plants and their pollinators. This book is based lovingly on their work. Any factual errors, omissions, or misstatements remain entirely our own and are not necessarily the views of our respective institutions, collaborators, or affiliated environmental organizations. Nor do we endorse any products or services listed in the appendixes.

# The Forgotten Pollinators

## Remembering the Pollinators

It is late in the fall—nearly winter solstice—but because the day is sunny, it is not too late to find a yellow and black bumblebee working one of the last blooms in our garden this year. It alights on a wandlike branch of a desert fairy duster, and comes under the spell of its brilliant flowers, a patch of brightness in a landscape otherwise gone to seed. The fairy duster's stamens are as red as a clown's nose, and the bumblebee clambers over, around, and through them in a manner that is anything but bumbling. It is there to gather nectar, a sweet liquid diet that is increasingly scarce at this time of year. And yet, from the flower's perspective, the bumblebee itself is a scarce resource; it has produced plenty of nectar and pollen, but until that creature landed in its midst, its own reproduction was in no way assured.

The brilliance and the showiness of flowers is but a visual reminder of the fact that pollinators are so often in scarce supply. It is worth their costly investments in advertising—petals serving as scented billboards—to attract insects to their midst. And not just any floral visitor will do; certain pollinators are more effective, and more allegiant, than others. The diversity of life in a place is not simply a random assortment of things; it is a fairly predictable set of organisms connected by certain ecological processes. Pollination services are among the most highly interactive processes involving both flowering plants and animals. When the pollinators foraging and nesting in wildland habitats provide pollination services to adjacent agricultural croplands, backyard gardens, or orchards, we say that these agricultural landscapes have benefited from the services offered by surrounding natural communities.

During the Persian Gulf War in the arid Middle East, one of us was sent out into another desert—the Sonoran Desert—to observe animal competition for scarce resources. In this case, the scarcest resource of the season was not fossil fuel in the form of crude oil, but nectar, the highly refined fuel used by hummingbirds. Along a stretch of dry streambed hardly the size of a football field, Costa's hummingbirds had been in residence all winter long, but they were now being invaded by other migratory hummingbirds arriving from Mexico. Each clump of chuparosa bushes was heavily laden with blood-red flowers, and each floral patch became a defensible territory.

Just before dawn, the whiz and whirr of the hummers' wingbeats would begin, and continue well into the day. Each would gain a foothold over a particular bush by chasing away other birds, or even butterflies. Then it would hover and flick its tongue into a tubular flower full of nectar. One hummingbird would make its way around a bush, darting toward one flower, then zigzagging up to another, then more, until it caught sight of an adjacent shrub laden with blooms. It would feed upon the floral sugars of the adjacent shrub for a while, then stop to roost on a perch atop the highest branch, resting between feeding excursions. And yet the rests would never last long, for another hummingbird would enter the scene, and a high-speed aerial chase or "dog fight" would begin once again.

The relative barrenness of the surrounding Sonoran Desert, and its paucity of bird life, underscores an essential but often unrecognized feature of that patch of chuparosa bushes. For migratory nectar feeders, it is an oasis in the desert, a nutrient-rich island in a sea of no-calorie sand. For the chuparosa bushes, hummingbirds are not the only pollinators available, but they are among the most allegiant and efficient ones; they seldom "waste" the chuparosa pollen sticking to their beaks and feathers on other kinds of flowering plants, as honeybees often do. Nor do the diminutive frenetic birds slit the floral throats of the chuparosa blossoms as do the nectar-robbing black giant desert carpenter bees, whose tongues are too short for legitimate front door entry. Eliminate all hummingbirds from this oasis, and the density of shrubs—even their very spacing—would no doubt change.

This is a book about one of the worlds' most vital processes linking plants and animals—a process that not only keeps us fed and clothed but feeds our domesticated animals and their wild cousins as well. Even more important, it keeps the verdant world, that delicate film of life around us known as the biosphere, running with endless cycles, feedback loops, and checks and balances. That ecological process is *pollination*—linking plants and animals. In fact, the range of animals active in moving pollen from one plant to another is bewildering in its diversity. In turn, many families of seed plants have diversified into their present array of species under the evolutionary influence of the myriad animal pollinators on this planet. And all these transactions between pollen-producing plants and pollen-moving animals make up a significant portion of what biological scientists are now calling *biodiversity*.

And yet, as the twentieth century nears its close, most North Americans lack any mental *image* for this "biological diversity" that scientists deem so important. Although this recently coined word has been splattered across headlines in innumerable newspapers, magazines, and radio and television broadcasts during the past decade, poll after poll confirms that few Americans understand (or care?) what ecologists and other scientists actually mean by biodiversity. Similar polls indicate that few Americans know that pollen plays a role in plant reproduction, for most of them regard it as a nuisance, an allergenic dust. And fewer still seem to know that the current rate of species loss constitutes a biodiversity crisis of unprecedented proportions. Scientists have barraged the public with mind-numbing numbers, species/area curves, equations, doomsday predictions. But they often fail to convey a sense of just how much we all depend on this flamboyant diversity of lifeforms, or how it is responsible for what we eat, drink, and wear. When people do finally hear of the biodiversity crisis, too often it sounds as though it is happening far away, in some exotic rainforest, and not in our own suburban backyards, our neighborhoods, our vegetable gardens, our agricultural croplands, in our supermarket produce department or at the local fast food burger, taco, or pizza joint.

But the truth of the matter is this: a pollination crisis has now become obvious in rural as well as urban settings not only in North America but on other continents as well. It is not merely an issue for rainforest activists, vegetarians, or beekeepers. It is an issue that can help us find common ground between farmer and forest ecologist, between bee-keeper and Mayan shaman, between organic gardener, pest control operator, and bat conservationist. To find such common ground, however, many of us will require more than merely reading agricultural statistics and estimates of species diversity. We will need tales, fragrances, tastes, and images that inform us about how the world works and what is at stake if we simply ignore the needs of pollinators and the habitats where they make a living. And so the two of us will be telling you stories interspersed with explanations of global trends, while Paul Mirocha will be offering new and visually arresting images to go along with our stories. In this way, perhaps, the principles of pollination ecology will be understood more palpably, and the loss of species will be considered not as a biological numbers game, but as the heedless destruction of other lives that have enriched this fragile earth. In these ways, we think, each of us is attempting to remember a life.

GARY REMEMBERS:

Often it is hard to discern when or how a mutualism began its development, but in this case, I believe it was in 1984. That was when Steve and I took our first lengthy field trip together—to a place where the desert meets the tropics, where Western civilization meets Mesoamerican civilization, where modern agribusiness meets ancient indigenous agriculture. As Steve and I drove into the foothills of the Sierra Madre—the Mother Mountains of Mexico—the seeds of a later collaboration were sown. As we traveled the narrow corridor that twisted its way deeper and deeper into the wilds of the sierras, we learned of our complementary interests and skills. Those mountains—and the juxtapositions they contained—gave rise to this book, which is as much about the conservation of ecological processes as it is about the health of our farms and wildlands.

We happened upon a place—one of the few still left in North America—where pollinators continued to connect the gene pools of wild fieldside plants with those of native field crops. It was a place where squash and gourd bees with ancient allegiances to native plants were beginning to be outcompeted by introduced honeybees. It was also a place where lesser long-nosed bats, the pollinators of century plants used for local mescal-making, were being dynamited out of roosting caves due to a misplaced fear of recently arrived vampire bats. As it turned out, these junctures form the seams that run through the stories you are about to read.

But I am getting ahead of the story. Earlier in 1984, while finishing the fieldwork for a book with Paul Mirocha, *Gathering the Desert,* I had begun to hear stories from Native Americans in Arizona, stories that puzzled me. They suggested that gene flow between their farm crops and local weeds had diversified their crops within recent history, as re-counted in their oral traditions. In particular, traditional O'odham farmers of southern Arizona claimed that their cultivated squashes oc-casionally became bitter due to contact with the fetid-smelling wild gourds that grew nearby. They also said that some of their chile peppers had become too hot to eat, presumably under the influence of the fiery chiltepines that grew spontaneously in nearby canyons. Here seemed to be evidence of what the great ethnobotanist Edgar Anderson called *in-trogressive hybridization*—the continual flow of genes between two closely related plants that sometimes results in new fieldside or roadside weeds.

Curiously, though, the wild species that were cross-compatible with peppers and squashes seldom grew adjacent to Indian fields in Ari-zona. This made me suspect that the oral traditions were coming from areas of Mexico just south of Arizona, where my O'odham friends have distant cousins who live where wild gourds and chiles are om-nipresent. Furthermore, the elderly Indian farmers in Arizona could not recall what pollinators came to visit native squashes and peppers, so it was hard to predict whether pollen could actually be carried sub-stantial distances between field crops and nearby patches of their wild relatives.

Sometimes a great notion like this—my hunch about gene flow—first strikes me as a little wacky and unscientific, so I knew I had to be careful when trying to document where gene flow had persisted in In-dian fields. My first inclination was to call fellow Tucsonan Steve Buch-mann, for he was a respected pollination ecologist who had the tools and methodologies to confirm whether or not chiles were indeed get-ting hotter and squashes more bitter. Fortunately, Steve said he was in-terested in the problem. He soon agreed to help me set up experiments to determine if gene flow between wild plants and crops still persisted in Mexican Indian fields.

And so Steve and I headed south across the desert border to Onavas in the state of Sonora one hot, humid day in August. There, where the Rio Yaqui snakes its way between the emerald foothills of the Sierra Madre, we came upon some of the northernmost stands of deciduous tropical forest where it interlaces with Sonoran Desert scrub. The Madrean foothills were covered with giant columnar cacti, umbrella-like canopies of thorny legume shrubs, giant kapoks, morning glory trees. Beneath them, on the verdant floodplain, we came upon a modest patch of field crops tended by an O'odham elder, Don Pedro Estrella.

Around the first of August, Don Pedro and his mescal-making son had taken time out from their other labors to plant a patch of maize, squash, and other crops where they had access to irrigation. By the time we arrived, the squash vines had just begun to emerge from the desert soil, but in the adjacent scrublands, gourds were already twining wildly over shrubs and stumps. Steve and I walked around the field, up to our thighs in the rank, teeming growth of *chichicoyota* gourds. Steve came up to a gourd plant and stood there a moment, aerial net held poised for a quick swipe at passing insects doing aerobatic loops.

Suddenly he swooped the net down over a big yellow-orange gourd blossom and captured a large solitary bee. Moving it to a jar, he quickly identified the robust orange bee as *Xenoglossa strenua,* one that pollinates only squashes and their gourd relatives in the genus *Cucurbita.* Here—as our research later documented in finer detail—was an animal that had coevolved with the squash and gourd family and was still serving as a bridge between wild and domesticated species in the cucurbit family. Without such mobile bee go-betweens, the plants themselves had no way of exchanging genes or ensuring a replacement generation of seedlings. And that, of course, is the crux of the matter and the purpose for our sojourn to those Sonoran farmlands.

Most terrestrial plants—save tumbleweeds and a few other vagrants—are profoundly rooted. That is to say, they must face the consequences of being *sessile* organisms, firmly stuck in one place. Nevertheless, a great variety of plants require the dependable movement of pollen from others of their kind in order to set seeds. Their need for pollen transport to ensure cross-fertilization with fruits containing germinable seeds is the norm among gourds, squash plants, and their tamed cousins the pumpkins. Thus someone (insect, bird, mammal) or something (wind, water, gravity) must carry that pollen from one plant to its kin. If a capricious gust of wind or a hungry sweat bee picks up the pollen but carries it to a different species of flower, the entire sexual exchange becomes literally fruitless.

At another level, genetic exchange between plants of a sizable plant population may allow them to escape the effects of harmful mutations associated with inbreeding. In the case of cultivated squashes, some of them no longer have natural resistance to powdery mildew and other crop diseases. By finding and transferring resistant genes in feral gourds to hybrid squash varieties, however, modern crop breeders have produced the powdery-mildew-resistant varieties that bless your dining room table or allow a fancifully carved jack-o'-lantern to scare little boys and ghouls on Halloween.

Sessile beings that they are, many terrestrial plants have evolved chemical and physical attractants which increase the probability that certain animals will find their flowers and carry away their pollen to others of their own kind. Both the go-betweens and the flowering plants themselves can thus be called *mutualists*. In exchange for transferring pollen from one plant to its nearby relatives, mutualistic pollinators obtain food, shelter, chemicals, or mating grounds in and around the flowers. (As we shall see later, these exchanges are not always fair trades.) Fair or not, this intrafloral commerce by the birds and bees is what makes the living world go round on its reproductive cycle. And yet this fact—fundamental to our own food getting—is easily forgotten by most city folks who don't gather their food daily from the vine but hunt down their polished fruits and shrink-wrapped vegetables at the local supermarket.

*S*TEVE REMEMBERS:
After our initial visit to establish the pollination studies in Onavas, Gary worked hard for two years to document gene flow between crops and their wild relatives. Other talented scientists joined us in the field or lab: Laura Merrick, Josh Kohn, Cindy Baker, Amadeo Rea, Ellen Ordway, Charles Shipman, Fernando Loaisa-Figueroa, and Bettina

Martin. We set up fluorescent dye-tracking of "pollen flow" between squash and gourd flowers. We identified native bee pollinators. We rated fruit for their "hybrid" characteristics. We ran electrophoretic analyses of genetic variation in our laboratories. And we interviewed local farmers and consumers. Some of the local Sonoran townspeople, of course, were puzzled by Gary and his ever-changing crew.

To fully document cross-pollination and hybridization in the field, a variety of these methods must be used in an integrated way. Yet some produce immediate results, while others take months or even years to correctly analyze and interpret. Eventually we documented one-way flow of gourd pollen to squash blossoms mediated by native bees and, to a lesser extent, outcrossing between cultivated chile peppers and wild chiltepines mediated by sweat bees.

But it was when I first had that big orange gourd bee buzzing in my net that I realized there was a valuable lesson to be learned here: Unless we could interpret how pollinators entered the picture, it was hard to understand the origin of a crop plant or weed, let alone assess the health of an agricultural ecosystem. Two decades ago, our colleague Peter Kevan from Guelph University in Ontario wrote that "the often un-known but undoubtedly important interrelationships of pollinators and plants constitute a serious void" in our understanding of both agricultural and natural communities. Much work has been accomplished since Peter warned us to be humble about the meager magnitude of our knowledge, but his assertion remains true to this day. Precise knowledge of pollination ecology remains the weakest link in our efforts to keep endangered plants from suffering further declines. It is also a weak link in efforts to sustain agricultural productivity to feed the world's burgeoning human population.

The same day I captured gourd bees in my insect net near Onavas, I was humbled by the implications of collecting another set of bees in the tropical forests nearby. I put up some scent blotters on tree trunks that soon attracted a number of brilliantly metallic green, long-tongued eu-glossine bees of the same genera I had studied on rare orchids in Panama. Here, hundreds of miles north of where they were supposed to be, according to distribution maps, I had obtained new records of these "orchid bees" for the state of Sonora. It reminded me of how little we knew about the distributions of some of the most spectacular pollinators of the Americas. I soon realized that we knew even less about human disruption of their habitats.

As we were leaving Onavas, Gary mentioned that many new white hive boxes filled with honeybees had arrived since his last visit to the pueblo. They were marked with the names of bee distribution points from all over Mexico. From my previous work with the U.S. Department of Agriculture, I knew that international development organizations had been actively promoting beekeeping as an economic pursuit in rural Mexico. Yet I didn't realize how ubiquitous intensive honeybee management and honey production had become, even in the remote reaches of the Mexican sierras. What, I wondered, was honeybee competition for pollen and nectar doing to the myriad kinds of native bees or other pollinating insects in the region?

The transfer of bees from one part of the country to the next worried me too. What if the new bees carried parasitic mites or diseases that had been causing declines in other regions? I worried that honeybees might be destined for a boom and bust cycle that could leave pollination services up to the already declining native bee pollinators remaining in the area.

As we drove northward toward the U.S. border in Nogales, Arizona, I realized that honeybee competition was but one of many potential threats disrupting the old relationships between native plants, crops, and pollinators in the region. We witnessed immense graders and tractors scraping away the ancient Indian fields there, releveling and consolidating irrigated plots to develop vast plantings of cotton and safflower. We saw campesinos casually toting backpack sprayers of pesticides, often misused to eliminate most of the target and nontarget insects in their wake. We saw roadsides recently sprayed with herbicides to open up the native vegetation for planting exotic African pasture grasses for cattle ranching. As we arrived back in Arizona, we passed field after field that was clean-cultivated and even laser-leveled for irrigation demands, a sterile landscape clear up to the highway edges. There was little room left for native pollinators, especially ground-nesting bees, in such highly engineered agricultural landscapes. If land uses did not eventually change for the better, there would soon be few places left for a lone gourd vine or a solitary bee to raise its weary head.

And so this book about the world's "Forgotten Pollinators" is rooted in our mutual curiosity regarding not only plants but their pollinators, not only natural history but cultural history in past and present times. Although one of us is trained primarily as a bee and floral biologist, and the other as an economic botanist, we both enjoy being in the field with scientists and naturalists of other persuasions. It is around the campfire, at the field station, in the remote study site, that such people often tell their most astonishing stories to one another, more charged with emotion and less cluttered with technical jargon than the tales they tell when back in the laboratory or lecture hall. We hope that our stories from the field make the critical topic of pollination ecology come alive for you and, perhaps, plant some seeds in fertile soil.

In many ways, however, this book is about communities where pollination ecology has been neglected. Along our migratory route from the United States to Mexico and back, we have compiled a surprisingly large body of scientific evidence documenting disrupted interactions between plants and pollinators, the diminished seed set being found among rare plants and commercial crops, and the decline in population sizes of animal pollinators. We have been reminded that the biotic interactions essential to the survival of most endangered species are poorly understood and inadequately monitored. In one region of the United States, for instance, the pollinators were known for only one in every fifteen endangered plants granted federal protection. Yet if the pollinators of these species remain unknown and unprotected, what are the chances that the plants themselves will continue to reproduce? In many cases, pollinator protection will entail reforms in international policy, for many migratory pollinators range between countries with very different conservation attitudes and capabilities.

We have been reminded, too, that early warning signals from disrupted wild habitats foreshadow future problems that we will be facing on food-producing arable lands. Much of the best farmland near large cities has already disappeared under houses, parking lots, fast food restaurants, and shopping malls. The conservation biology issues we present here are not simply esoteric concerns relevant only to middle-class bird-watchers and bug-netters. These issues should strike a chord in every person who cares about where our food comes from and whether it is wholesome to eat. After all, one in every three mouthfuls you swallow is prepared from plants pollinated by animals.

For this reason, we can only hope that this book will be read by as many farmers, gardeners, grocers, gourmet cooks, land managers, policymakers, and orchard growers as it will be read by scientists, students, teachers, and amateur naturalists. Chapters 1 to 5 introduce the fundamental principles of pollination ecology through "parables," classic case studies, explaining plants, pollinators, and pollination processes. Chapters 6 to 8 explain how these processes have been disrupted. Chapters 9 to 12 address the question of agriculture and landscape restoration. In addition to the appendixes giving useful details of pollination ecology and sources, there is a bibliography and a glossary of technical terms. In the text, these terms are printed in *italic type*.

*A lesser long-nosed bat* (Leptonycteris curasoae) *visits the blossoms of a giant columnar cactus — the saguaro* (Carnegiea gigantea), *Arizona's state flower — on a hot summer's night in the Sonoran Desert of Arizona.*

# CHAPTER 1

# Silent Springs and Fruitless Falls

## The Impending Pollination Crisis

S TEVE REMEMBERS:
The Virgin River basin of southwestern Utah is a place of rarities, from towering crimson canyon walls and Joshua trees to a host of low-lying, lesser-known living treasures. It was my first time to the Virgin, but Gary had made pilgrimages here before, collecting rare wild sunflowers to be used in breeding disease resistance into commercial hybrid sunflower crops. This trip, however, we came to the Virgin not for its sunflowers but to stalk a rare diminutive poppy. That poppy grew on gypsum-laced bluffs and hummocks too bleak for much of anything else to grow, but it had one local animal associated with it that we hoped to meet—a bee as rare as the plant itself.

We had flown into the virtual-neon-baking-hot reality of Las Vegas one morning in late spring. There we rented a vehicle to take us two hours to the northwest, into the higher reaches of the Virgin watershed where the crimson cliffs of the Chinle Formation enclosed sandy valleys. Below us, in one valley, a patchwork of greener-than-green alfalfa fields juxtaposed themselves against the badlands of pink dunes, cocoa-colored mudstones, and pale gray, bleached-out gypsum hummocks. From the very base of the nearby cliffs, low-growing familiar shrubs of pungent creosote, snakeweed, and saltbush were sparsely dotted across the flats, their roots competing for the meager rainfall of that semi-arid land. It seemed like an unlikely spot for us to look for one of the most endangered pairs of plant and animal in all of North America, but our hunt was on.

The poppy in question has a lovely, ivory-colored blossom, but a name that few people have heard: *Arctomecon humilis*. One animal associate of the "bearclaw" poppy is recognized by even fewer people, for it is a bee recently described as a new species: *Perdita meconis*. The story of their barely surviving together in a not-so-virgin habitat is a reminder of how little we know about the rarest-of-the-rare, even in a nation that spends more on environmental monitoring and protection than any other country in the world.

It could be that the bearclaw poppy was never widespread nor naturally abundant. Nevertheless, it has become exceedingly rare over the last century of habitat degradation, and the last decade of frequent drought. It was listed as a federally endangered species in 1979. Soon after that, during a five-year dry spell, nearly every seed-producing poppy plant died off, and hardly any new seedlings were recruited to replace them. Most of these perennials live a scant five years at the most, but far less than that when stressed. By the late 1980s, bearclaw poppy numbers had become so thinned that the conservationists who knew it best could only predict that it would be extinct by the year 2000.

We arrived at the end of a wet spring, so the prognosis did not look so bleak to us. Still, Gary and I wandered up and down the mudstone and gypsum hummocks of the Moenkopi Formation for well over an hour before I stumbled across the first flowering poppy. Excitedly I yelled for Gary to come over, for I doubted that he could see it from any distance away: it was barely 6 inches tall, with waxy blue leaves. The flowers bore velvety white petals and a central mass of bright orange-yellow stamens, looking like a miniature fried egg, sunny-side up.

We looked around, and realized that we were in a broad arc-like swath of a hundred poppies or more spread over 50 yards or so. Most of

*The diminutive bee approaching the petal is* Perdita meconis. *In a very small number of sites—like this one in southern Utah—it collects pollen and nectar from the rare dwarf bearclaw poppy* (Arctomecon humilis) *and in the process pollinates it.*

the plants had finished flowering, but the bulk of them had bloomed later than usual, so we were lucky to see quite a few blossoms. In fact, we were lucky to see any poppies at all. Cattle hoofprints and off-road vehicle tracks wandered right through the largest patch of poppies, despite a plethora of formidable signs posted nearby warning that the area was closed to traffic.

The abundant rains had encouraged high fruit production for the coming season, but when I knelt to examine the fruit, I realized that good weather was not enough to ensure that the fruits would contain a full complement of seeds. Each partially opened fruit looked like a miniature Easter basket, complete with handle, holding the seeds until they could germinate with the summer rains. A fully pollinated fruit might have over 30 thick, shiny black seeds in it. Yet many of the fruits that I scanned were already shrunken from lack of pollination, or perhaps from abortion resulting from the poor nutritional condition of the mother plant.

We took no seeds from the site, but we could easily count their numbers in the earliest-maturing fruits. They had just begun to dehisce, drying and splitting in fissures. A few fruits had 25 to 30 seeds in them, but most had far fewer. I rattled off my counts to Gary: 23, 20, 12, 14, 7, 18, 1, 4, and 11 seeds per fruit. Either many of the plants lacked the nutritional reserves to mature all the ovules into viable seeds in their fruits, or their flowers had not been pollinated to begin with: the average fruit held less than half the maximum seed complement it could produce under optimal conditions.

The only potential pollinators we saw during our dawn and dusk visits to the poppies were introduced honeybees, likely from a nearby apiary. They were busy on evening primrose and buckwheat blossoms, and for the most part, they left the bearclaw blossoms alone. Native pollinators may have been more active on this site earlier in the season, but they were nowhere to be seen while we were present over a two-day period.

Fortunately, in 1988, a group of entomologists from Utah State University had better luck at the site than we did. In May of that year, Vince Tepedino, a USDA-ARS research entomologist stationed in Logan, Utah, encouraged his student assistant Bonnie Snow to collect at the site. Bonnie swept up a species of solitary bee hitherto undescribed by entomologists, one that had evaded notice for more than a century after the poppies themselves had been described by biologists.

In fact, the bee had been collected just once before, on another kind of poppy, in the Kelso Dunes of eastern California well over a hundred miles west of the Virgin River. Terry Griswold was the collector of the Kelso Dunes bee, and when he saw that Bonnie Snow brought the same kind into his lab, he realized that these specimens were unlike any other from the region. Griswold named the new species *Perdita meconis*. His colleague Vince Tepedino soon went back out into the field to learn more about the relationship between the poppy-loving solitary bee and the endangered poppy.

Tepedino found that in the Virgin watershed, the poppy-loving bee has an unswerving allegiance to the bearclaw poppy despite the local availability of a wide variety of other plants blooming within its flight range. The bee has not even been found on other kinds of poppies closer than the dunes at Kelso. Over the entire time it has been observed, this solitary bee has maintained its fidelity to just these two poppy species, which have yet to be found growing together.

In short, Tepedino, Griswold, and colleagues have documented what theorists have assumed to be a highly improbable occurrence in the natural world: a pollinator that, on at least one site, specializes exclusively in visiting a rare plant, one with an extremely restricted distribution. A bee that visits only one kind of flower is called a *monolege;* the poppy-loving bee, however, is more precisely a highly restricted *oligolege,* a specialist on a small set of closely related flowers, one at each site. There are only a handful of well-documented cases in North America of truly *monolectic* bees, ones that associate with just a single source of pollen. Among them are *Trachusa larreae,* which is exclusively dependent on creosote bush in the desert Southwest; *Cemolobus ipomoaeae,* which is hosted exclusively by a single morning glory; and *Hesperapis oraria,* which visits but one dune-dwelling sunflower inhabiting a 200-mile stretch of Gulf Coast plains between New Orleans and the Florida Panhandle.

Such extreme specialization is clearly not the norm in the natural world, but it is the kind of exception that proves the rule: the fates of many plants and their associated animals are, at one level or another, intertwined. This is sometimes called the "Dodo Principle of Linked Extinctions," for when the dodo was extirpated on the island of Mauritius 300 years ago, a tree dependent on the Dodo began to decline in numbers. Why? Some scientists argue that its seeds require passage through the digestive tract of the extinct bird—or one with an equally impressive ability to crack open hard seedcoats—in order to germinate. In the case of monolectic bees and their host plants, the plant's decline would be expected to trigger a marked response in its associated animal population.

Theorists have long argued that such restricted relationships hardly make evolutionary sense: why would a pollinator confine its foraging to a single resource, particularly if that resource was naturally scarce to begin with? In the case of the bearclaw poppy, it may have never had a wide range, but it has undergone a dramatic local decline in numbers. Apparently the poppy-loving bee has not read current theories of optimal foraging, or it would have abandoned the poppy and switched to other floral resources.

Instead, as Tepedino has learned, the poppy requires a bee to serve as a vector for pollination, since cross-pollination produces significantly more seeds. This bee visits multiple flowers of the poppy during a single foraging bout, regularly contacting the stigma of each blossom, and has been observed repeatedly moving between the plants, resulting in cross-pollination. In short, the poppy-loving bee is not a casual or promiscuous floral visitor. Its behavior and form have evolved in such a manner that it has become a consistently effective pollinator of this rare plant.

Sitting in his office in Logan, Utah, one day, Tepedino proposed to us that "this bee should perhaps be listed as an endangered species just as the poppy is, since it is known from so few places." If we duly acknowledge how human pressures have already exacerbated problems posed by their natural rarity, this would become the first case in the continental United States where both a plant and one of its key pollinators would both be federally protected. This is not to say that the poppy itself would face extinction if the bee were extirpated from the Virgin watershed, for two other bees also visit the bearclaw poppy. But the three bees visit sequentially, in waves, over the poppy's flowering season. The first to appear is a fairly common native bee known by the melodious name of *Synhalonia quadricincta;* next comes the rare poppy-loving bee, *Perdita meconis;* last, and "certainly least" says Tepedino, comes the European honeybee, an outsider.

We had come to the Land of the Bearclaw Poppy so late in its flowering season that it was not even being visited by many of the managed or feral honeybees in the area. A few miles to the north of us, where poppies formerly grew, housing developments had taken their place. We can only guess at the former range of the bees themselves. Years ago, Stan Welsh, the dean of Utah's botanists, wrote that the bearclaw poppies "should be regarded as national prizes, as jewels of great price, and protected for future generations." What he did not know is that future generations of the poppy-loving bee will not survive unless these floral prizes remain alive and well within their native habitat.

Consider what might occur if, by happenstance, a global or local warming trend extends the duration of drought past seven years in length—the presumed maximum lifespan of a cohort of poppies. The amount of pollen and nectar available to feed the bees will be too small to nourish their population. Should moist conditions return, dormant poppy seeds may later germinate out of the Moenkopi Formation's gypsum soils, but they may not find even one bee left to pollinate their flowers.

It is conceivable that introduced honeybees or other (insect) generalists might provide some pollination services to the poppy. Nevertheless, a millennia-old relationship between poppy-loving bee and bearclaw poppy will have ceased to function. And almost certainly, the poppy's survival will be at greater risk, for its backup pollinator—the once-abundant honeybee—is now suffering dramatic population declines throughout North America.

There is good news and bad news when it comes to placing this parable in some sort of global context. The good news is that monolectic and oligolectic bees comprise a rather small percentage of all known pollinators. As we hinted earlier, few plants dance with just one partner. The bad news is that linked extinctions should not be the only topic of concern for conservationists who care about endangered plants or their rare pollinators. A formerly diverse natural community may become impoverished even when one or both partners in pollination decline, but do not become globally extinct.

Auburn University botanist Robert Boyd has recently written: "The most prudent approach to conservation of [a locally rare plant] would be to preserve and manage not solely [the plant] but as many components of the ecosystem as possible. This would likely include the pollinators, seed dispersers, and other organisms that perform important roles in the plant's life cycle." But such an ecosystem approach might be costly, for it assumes that we know a lot about the dynamics in interactions between plants and animals. In fact, we are only beginning to understand even the most obvious relationships between flowers and their pollinators.

# GARY REMEMBERS:

In some cases, it is not the plant population that first begins to dwindle, but the animal with which it may have coevolved as they reciprocally adapt to one another over the centuries and millennia. I have witnessed such a situation several times over the last few years, a thousand miles away from the Mohave Desert in the heart of Mexico City. I cannot imagine a habitat more distinct from that of the poppy's as I found in the Pedregal, an overgrown lava field located on the city-like campus of UNAM, among the largest universities in the world. There, within easy walking distance of some of Mexico's finest laboratories, ecologist Luis Eguiarte has taken me to a pollinator monitoring site that he established with our mutual friend Alberto Burquez in 1982.

On many nights between 1982 and 1994, Luis and Alberto have sat out among some of their favorite flowering plants from sunset through early evening. They have tallied the numbers of nectar-foraging bouts made by the lesser long-nosed bat, a species presumed to be declining in central Mexico. The flower that Luis and Alberto favor is a distant relative to mescal-producing century plants. It is a small, rosette-forming succulent with no common English name, known only as *Manfreda brachystachya*. I have nicknamed the plant "Señora Manfreda," although I am sure Luis and Alberto do not condone this anthropomorphizing.

Señora Manfreda, like many of her mescal-producing kin, emits a musky scent and abundant nectar within her pale tubular flowers. The flowers themselves are stacked laterally along a 4 to 6-foot-tall stalk and arranged for easy access by nectar-feeding bats. Although other kinds of bats as well as hawkmoths and hummingbirds visit Señora Manfreda, the lesser long-nosed bat remains her most common visitor. This bat is not just another nectar-sucking pretty face; it is a highly effective pollinator.

And pollinate they do. Luis has shown me how he tags each flower stalk that has been visited, so he can later correlate bat visitation frequencies with the number of fruits set by each plant. I can recall the surprise in his voice when he recounted the discovery of a 75 percent reduction in fruit set between 1982 and 1985: "At first we thought we were witnessing a permanent decline in fruit set. Why? Then we realized that our first season's visitation by bats had been much higher than during our second season in 1985." Luis and Alberto went on to demonstrate that declines in fruit set directly reflected declines in bat visitation rates to flowers, even though other floral visitors—hawkmoths and hummingbirds—were still present. They also noted that in a botanical garden not far from their site, century plants that had received 100 bat visits a night in 1982 were down to 20 visits a night in 1985.

I probed Luis for more detail. "After a few years' worth of more data," he said, "I realized that we were witnessing fluctuations—but not necessarily unidirectional declines—in both the local abundance of bats and the fruiting of plants." Luis sat down and showed me chart after chart documenting a tight correlation between two phenomena: visits by pollinators and fruit set. When bats visited the plants frequently, fruit set was high; when they were nearly absent, fruit set was low. Evict the bats from the Pedregal, the data predicted, and rates of fruit set would take a corresponding dip.

As I looked beyond Luis, and palpably felt the pressures of the burgeoning metropolis, I could not help but wonder how long the bats would continue to take up residence in one of the biggest, smoggiest cities in the world. One of the plant's study sites has already been lost to new building construction on campus. The university's bat experts have no idea where the bats can still find a safe sanctuary in or near the Pedregal. Although not yet on an irreversible downhill slide, the fruit set of Mexico City's "Señora Manfreda" will ultimately depend on how many bats survive urban nightlife. Other nectar-feeders may still forage for nectar in the Pedregal each night, but should the bats vanish from the scene, Luis and Alberto will have far fewer fruits to count each fall.

It has been well over 30 years since Rachel Carson predicted a silent spring, one devoid of the chorus of insect-feeding birds, one where "no bees droned among the blossoms." That prophecy was heard far and wide, and perhaps more than any other of the last half century, it changed the way farmers, wildlife managers, and policymakers perceived "environmental protection." Yet Rachel Carson also predicted fruitless falls, autumns in which "there was no pollination and there would be no fruit."

Carson suggested that fruitless falls would become more commonplace in the American countryside for two reasons. First, she said, "a bee may carry poisonous nectar back to its hive and presently produce poisonous honey." This prediction proved true, and considerable efforts have been made to reduce domestic honeybee poisonings by herbicides and pesticides. However, the same effort has not been diligently extended to protect *wild* pollinators from toxic chemical exposures, direct and indirect, on farms and wildlands. Second, she observed that "many herbs, shrubs, and trees of forests depend upon native insects for their reproduction; without these plants many wild animals and range stock would find little food. Now clean cultivation and the chemical destruction of hedgerows and weeds are eliminating the last sanctuaries of these pollinating insects and breaking the threads that bind life to life."

Of all of Carson's commentaries, this is perhaps the one that has been least heeded or understood: that habitats are being fragmented by physical destruction and chemical disruption of their biota. The effects of such habitat fragmentation on pollinators and seed set have recently been documented in habitats as different from one another as Iowa's tallgrass prairie pockets and Argentina's dry "Chaco Serrano" scrublands. Once again, Carson's superlative intuitions were right on track and eerily futuristic.

It now appears that the majority of plants studied to date show evidence of natural *pollinator limitation*. That is to say, under natural conditions, 62 percent of some 258 kinds of plants studied in detail suffer limited fruit set from too few visits by effective pollinators. If this condition is the norm in the natural world, to what extent is the regeneration of plants jeopardized by human disruption of the interactions between plants and their pollinators?

In Iowa, where tallgrass prairie once covered 5 million acres, only 200 acres of virgin prairie remains intact. Prairie ecologist Stephen Hendrix has found extremely low seed set of three prairie wildflowers in the smallest, most isolated fragments of this remaining prairie. It is not that bees and butterflies no longer find a single phlox, purple prairie clover, or lead plant attractive. The pollinators have simply had to abandon the smallest patches of these plants because they lack sufficient rewards to sustain the animals' populations on these remnant habitat islands.

Two pollination ecologists, Peter Feinsinger and Marcelo Aizen, found much the same trends on a community-level scale in dry subtropical "Chaco Serrano" scrublands in northwestern Argentina. There habitat fragmentation over the last three decades has resulted in declines of native pollinators such as solitary bees; these declines have reduced the reproductive success of at least 16 different tree species. Fragments of 5 acres or less simply could not sustain sufficient numbers of wild pollinators, so that domestic honeybees (in this case, Africanized bees) have come to dominate the ranks of the pollinators.

According to botanists Subodh Jain and J. M. Oelsen, habitat fragmentation is just as much a "life and death issue" for cross-pollinating plants as pesticide accumulation is for insectivorous birds. In 1979, Vince Tepedino predicted that regions harboring high proportions of specialized pollinators or obligately outcrossing plants would soon be subject to high rates of species loss. In a recent interview, Tepedino sadly told us that his prediction may have been more accurate than he wished: "I was amazed that so many recently reported case studies fit with the predictions of that ancient paper. We're now realizing that generalist pollinators such as sweat bees are the only ones left to do much of the pollination for cacti and other rare plants."

For as much as Carson and Tepedino may sound like they both made their living as Prophets of Doom, there is one even more fundamental characteristic the two of them share: a love of inquiry into the natural history of relationships. And if that love led both of them to grieve for the extinction of certain relationships they considered vital to the living world, they also celebrated the vitality of extant relationships between plant and animal, human and other. To gain a sense of what is at risk, we too must first explore the diversity and complexity of intact relationships. That exploration will take us from the Virgin River watershed in Utah to the "orphan islands" of the Galápagos Archipelago, but it will never leave this underlying theme: a biologically rich place is rich in relationships as well as in species. Conversely: the loss of biodiversity is always more than the simple loss of species; it is also the extinction of ecological relationships.

Flowers offer their pollinators a strikingly
diverse array of shapes, colors, and scents
to lure them to hidden rewards. Depicted
here are 13 floral designs picked at
random to illustrate their rich abundance
of forms.

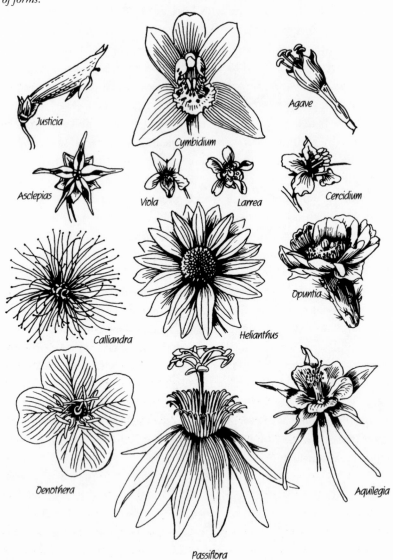

# CHAPTER 2

# Flowers

*Waiting for Their Ships to Come In*

G ARY REMEMBERS:

It did not matter how much I had read about the Galápagos Islands before arriving in the famous archipelago. Romantic notions kept floating
through my head as I set foot on their volcanic shores for the first time.
I had come with a group of students and professors to collectively study
the biogeographic patterns across the chain of 45 isles made famous by
Darwin, but the patterns already set up in my head were quickly reduced to chaos. My journey there was in 1973, long before Las Encan-
tadas became a mecca for ecotourists. Today the islands are under siege
by a yearly onslaught of over 25,000 camera-toting turistas, and plans
for luxury hotels are under way.

My focus was on the plants that colonized the coastlines of the Galápagos, but I was surprised to see how barren the land actually was. As I
clambered from the *Cristobal Carrier* to the Ecuadorian Park Service
dock, I scanned the glistening black, tide-splashed lava flows on Isla Fer-
nandina. There I was reminded of how easy my journey had been

compared to those taken by seeds, tubers, insects, birds, and reptiles when they "won the lottery" of long-distance chance dispersal from mainland South America outward to the archipelago. Somehow, they had arrived on these distant shores to establish reproducing populations.

I had assumed that because the Galápagos were so near the equator, each of the islands would be its own peculiar yet verdant *tropical* paradise. I imagined carpeted coastlines, lush with large, gaudy flowers, visited by all sorts of brilliant butterflies and hummingbirds. No matter how many times I had read about the drying effects of the Humboldt Current—or the difficulties of dispersal from the nearest continental landmass 500 miles away—I was still shocked by the bleaker realities. No wonder Herman Melville called them Los Huerfanos—"the Orphan Islands"—for their depauperate flora gave most of the islands the feeling of being inhabited solely by waifs, the castaways of the botanical world.

Instead of finding the rainbow of floral hues that I associated with tropical wildflowers, I saw the same small white and yellow blooms again and again. In place of lanky, intricate blossoms highly specialized for visits by discriminating animal pollinators—flowers like the orchids and birds-of-paradise I had seen in the lowland tropical forests along the Ecuadorian coast—I saw the tiny cups, bowls, and bells of generalists open to anyone. Alongside them were the pendant catkins of wind-pollinated colonizers—no prizewinning horticultural show worth writing home about to the local florist. I sniffed the air at Punta Espinosa without catching the waft of any perfumed fragrances from the floral world: my nostrils captured only the sting of the salt-saturated sea breezes.

We had landed against a tenuous strand of mangroves full of raucous pelicans. A green sea turtle was stranded in a lava-rimmed pothole until the next high tide arrived; its helplessness captured our attention for a while. From there, as we climbed onto a massive flow of pitch-black *pahoehoe* and *a'a* lava less than two centuries old, I was struck by how few animals I saw coming to visit the flowers available on the scraggly shrubs, cacti, and sedges jutting up from tiny patches on sparse volcanic soil. I spotted spiders, snakes, and lizards, but nothing immediately in the way of hummingbirds, hawkmoths, or butterflies. Could it be, I wondered, that the plants here are "orphaned" in another sense? They appeared as though they lacked pollinators to connect one flowering plant with the nearest of its kind.

Pollinator diversity *is* remarkably low in the Galápagos. The pioneering annuals on the beaches nearby are wind-pollinated, like the goosefoots and dropseeds, or somewhat capable of self-pollination, as the wild tomatoes can be when pollinators are scarce. In addition to the tomatoes, a few of the wildflowers such as morning glories, ice plant, and cotton must rely heavily on Darwin's carpenter bee *(Xylocopa darwini),* the lone bee species native to the archipelago. Butterflies—which often favor selection for bright pink, yellow, orange, or blue flowers, sweet-scented and deeply tubed—are represented by only two native species in the Galápagos winged bestiary.

Surprisingly, the treelike prickly pears of the islands do not depend on solitary bees for pollination as do most of their continental counterparts. Instead, the probing bill of the ground cactus finch, *Geospiza scandens,* is responsible for moving much of the cactus pollen from one towering prickly pear tree to the next. Despite the fact that few flowering species in the Galápagos produce copious nectar, this single species of Darwin's finches has become a part-time nectarivore. Eagerly it draws on the meager nutritional rewards still evident in this prickly pear flower, rewards found more abundantly in its mainland ancestors. Such time-tried nectar production harks back to the evolution of their cactus ancestors in habitats where nectar would attract a plethora of solitary bees as visitors and pollinators of their blossoms.

The pollination of prickly pears on another island—Sea Horse Key off the western coast of Florida—has helped pollination ecologists articulate a major principle relevant to the conservation of biodiversity. This principle explains certain features of plants found in small, island-like habitats where pollinators may be too scarce to ensure adequate levels of cross-pollination. It predicts that there will be strong selection pressures exerted which enable plants to *self-pollinate*—that is, the pollen in a flower will be produced at a time appropriate for fertilizing the same flower from which it comes. Without the ability to self-pollinate, a plant will likely be eliminated from an island's flora after suffering from low reproductive success.

In 1980, Eugene Spears boated out to Sea Horse Key to search for differences between island and mainland populations of plants, especially in their interactions with native pollinators. As Spears puts it, he was curious about three issues: "Are pollinators less abundant and less predictable for the same plant species growing on islands relative to comparable mainland sites? Is reproductive success reduced in island populations relative to mainland populations? ... [And] if so, can this be attributed to pollinator limitation?" *Pollinator limitation* is the key phrase here. It is a term biologists use to signal that plants are not receiving visits from pollen-carrying animals frequently enough to ensure high seed set per fruit. For a plant to be pollinator-limited is paramount to being out of reach of potential mates.

Spears decided to study two plants simultaneously. The first was *Opuntia stricta,* a prickly pear cactus with simple, bowl-shaped flowers that produce abundant pollen but little nectar. The other was *Centrosema virginianum,* a perennial legume vine known as the butterfly pea, whose highly specialized petals tightly enclose its reproductive organs. Insect visitors must be able to land on the keel petals and manipulate the winglike petals if they are to expose the receptive stigma and transfer pollen to it. Both plants grow on the Florida peninsula as well as on nearby islands such as Sea Horse Key, only 5 miles out in the Gulf of Mexico.

Over the years, Spears found that there were generally fewer bees visiting the flowers of plants on Sea Horse Key than on the mainland. This was especially true for visits to prickly pear flowers, which received only a third to a fourth of the bees per hour that the mainland flowers received. Honeybees are not present on Sea Horse Key, but the solitary bees and colonial bumblebees that did reside there were much more erratic in visiting both butterfly peas and prickly pears.

Curiously, the bees present in the Keys carried pollen between flowers for shorter distances than the bees on the mainland. Most of the prickly pear pollen on Sea Horse Key was carried less than a yard from its flower of origin. On the mainland, however, a significant portion of pollen was regularly carried beyond 4 yards, from one prickly pear to the next. The limited dispersal of pollen on Sea Horse Key may have increased the probability of *inbreeding,* which can in some cases be harmful to plant populations. Sometimes inbreeding (especially to plants in dissimilar habitats) can eventually depress plant vigor and reproductive success in populations—hence the term "inbreeding depression." In other cases inbreeding leads to rapid diversification, or what ecologist Peter Feinsinger calls "inbreeding euphoria," evident in plants such as the Hawaiian silverswords. Inbreeding is one of several processes that may aggravate the genetic

*A bumblebee* (Bombus sp.) *about to visit the bilaterally symmetric flower of the garden pea* (Pisum sativum).

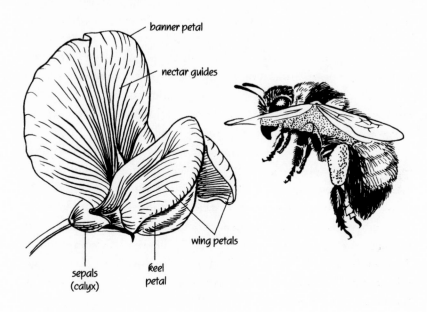

banner petal

nectar guides

wing petals

sepals
(calyx)

keel
petal

bottlenecks which already constrict most island-dwelling populations. Being able to mate only with your siblings and a few close neighbors often has its disadvantages.

The most telling differences between mainland and island plants became apparent when Spears hand-crossed butterfly peas, using masses of pollen ample enough to fertilize every ovule. He compared the fruit set from his handmade *outcrosses*—flowers fertilized with pollen brought in from distinctively different plants—with those from naturally pollinated flowers. Spears found that his pollen-dowsing of butterfly peas significantly increased their fruit production. But why? In 1981, bee visits to butterfly pea flowers were so infrequent that not one of the 40 flowers observed by Spears on Sea Horse Key set any fruit.

The paucity of pollinators in that isolated habitat was the main factor limiting the reproductive success—and ultimately the survival—of butterfly peas.

Prickly pear cactus flowers are generalists compared to butterfly peas. Because they can attract a variety of nonspecialized pollinators, cactus fruit set on Sea Horse Key was hardly different from that on the mainland. Yet the number of *seeds* per prickly pear fruit was far lower on the island than back on the mainland. This posed an intriguing question for Spears: "If differences among pollinator communities and in reproductive success among plant populations occurred over a five mile distance, how much more profound would the effects be in populations isolated by even greater distances?"

Spears concluded that "pollinator limitation may be an important selective force on plant communities"—especially for those of distant, oceanic islands. He proposed two ways that pollinator scarcity could shape vegetation. First, plants that were established in islandlike habitats might eventually gain independence from the specific animal pollinators on which their ancestors depended. They might accomplish this by adapting to self-fertilization or to wind pollination. They would then be able to set fruit in the absence of specialized pollinators—or any animals at all, for that matter.

Or, Spears suggested, "pollination scarcity could [act] as a selective filter, inhibiting the colonization of islands by plant species with floral structures and breeding systems that require specific animal pollinators." In other words, the paucity of pollen transporters may be the end of the line for any island-arriving plants that cannot get by without specialized pollinators. They will be unable to produce enough progeny to establish a permanent population on an island if they lack contact with those particular pollinators. Finally, Spears observed, "plants that colonize a new habitat risk leaving their usual pollinators behind." This is especially true for a plant with a flower as specialized as that of the butterfly pea. Lacking the bumblebees adept at "tripping" its flowers to reach its concealed pollen and nectar, the seed set of the butterfly pea plummeted.

There is a more graphic way of imagining Spears' "selective filter." Let's say you are a plant with a specialized flower. You find yourself stuck on an island large enough to offer suitable habitat for your potential mates, but without sufficient habitat or resources for the special pollinators that know how to get into your flowers. Your chance of passing on your genes—and producing more descendants to populate the island—will be nearly as low as that of two sailors (same sex) who are the only humans shipwrecked on the island. For most outcrossing species, a small desert island—whether off the coast of Florida or clustered with others in the Galápagos—can hardly be called paradise.

The story of Sea Horse Key is remarkably similar to one that Peter Feinsinger and Yan Linhart encountered a few years earlier during their studies of hummingbird-pollinated flowers on the islands of Trinidad and Tobago, miles off the South American coast. On the smaller of the two islands—Tobago—there were fewer species of hummingbirds. There was also greatly diminished fruit set on plants with specialized flowers compared to those with more open, generalized flowers accessible to a wide variety of animal visitors. Linhart and Feinsinger clearly documented the multifaceted risks of overdependence on a single pollinator, the same risks that worried Eugene Spears. As their hummingbird study suggests, "highly specialized plant-pollinator relationships are especially susceptible to perturbations of any sort, because any factor affecting the relative availability of either the plant or its pollinator necessarily affects both populations." At the same time, they recognized that this phenomenon was not restricted to hummingbirds and long-tubed flowers: "Plants that depend upon specific insects for pollination have greatly reduced seed-set if their pollinators are in low abundance, and such reductions can affect their geographic distribution."

The significance of their conclusions—or at least the magnitude of their applicability—depends on how widespread the phenomenon of specialized flowers is within the plant world. If generalized flowers are the norm—and blossoms with specialized flags, triggers, tubes, and tunnels the exceptions—it hardly matters whether a few bizarre blossoms get stranded on distant islands, unable to be serviced by the impoverished set of potential pollinators to be found there. But what if the myriad mainland flowers that we see and smell around us have radiated into such diverse forms in response to the diverse forms of their distinctive pollinators? If this is the case, specialized flowers may be a frequent occurrence. And the fragmentation of the world's habitats into islandlike patches would justify our timely concern, for specialized flowers are more than just pretty aberrations.

At the heart of this issue is our understanding of floral diversity. Simply put: why have flowers taken so many forms? To answer this question, we must recall what flowers are meant to do, and why such complex structures first evolved more than 100 million years ago. Flowers contain the plant's sexual organs and, within female organs, the potential seeds or gametes. The floral architecture must protect them from drying, scraping, freezing, burning, or browsing by herbivores. At the same time the flower's petals and sepals are protecting these organs, they must allow access so that pollen can come and go. More than that, the flower's shape must increase the probability that pollen will arrive on receptive stigmas at just the right time. Then the flower must continue to protect the stigmas while the pollen grains send down pollen tubes containing sex cells that migrate downward through stylar tissues to eventually fertilize the ovules. Finally, closed flowers often continue to protect developing seeds from the range of stresses that plague flowers and fruits prior to their opening, from insect predators to drying winds.

These functions are equally important for flowers that are animal-pollinated, wind-pollinated, water-pollinated, or self-fertile. These tasks are accomplished by very different floral forms, however, depending on whether wind, flying foxes, flies, hawkmoths, butterflies, hummingbirds, beetles, or bees are the agents of pollen dispersal. For seed plants, reliance on the wind to disperse pollen is the most ancient method for ensuring cross-pollination, which in turn reduces some of the problems associated with inbreeding. Soon after their Silurian conquest of the ancient landmasses more than 400 million years ago, early terrestrial plants took to releasing—almost like dust in the wind—copious quantities of spores and later true pollen, each grain poor in nutritional value, rather than investing in intricate floral structures, rewards, and fragrances to attract hungry animals. Without the aid of animal vectors, though, the airborne sexual particles of these land colonizers—clubmosses, horsetails, and numerous now-extinct groups—faced a daunting challenge of getting to "safe sites" on the reproductive organs of potential partners. Not only is the reliance on wind or water to transport gametes chancy, but it may even kill the sensitive sexual nuclei within the pollen grains.

*The parts of a flower are shown on this hummingbird-pollinated fuchsia blossom (Fuchsia sp.). Although flowers vary enormously in size and shape, they all have these basic parts, which have been molded through natural selection by their animal or abiotic pollinators.*

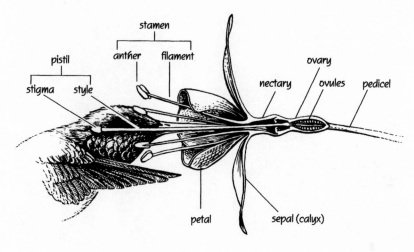

It may seem statistically improbable that a tiny grain of pollen or microscopic spore could be swept away by the winds, to float around indefinitely, only to land exactly "on target." And yet enough gametes won at this aerial lottery to have ensured the greening of the extensive swamplands of the Carboniferous. In fact, these wind-pollinated plants were so abundant and successful that their legacy of compressed organic detritus has fueled the modern world's industrial growth in the form of coal, crude oil, petroleum, peat, and natural gas. It is not hyperbole to claim that much of contemporary human culture—from automobiles to aircraft to greenhouse-ripened tomatoes—would have been impossible were it not for the sexual urges of ancient Carboniferous swamp dwellers, casting their gametes to the winds of chance.

Even today, thousands of plant species still rely on airborne transport of their spores and pollen grains—it is by no means an obsolete reproductive strategy. From the tiniest soil-hugging cryptogams to the tallest bamboos and stoutest trees on the planet, a great diversity of

*Enlarged pollen grains from a variety of flowering plant and conifer genera showing the great diversity in size and surface sculpturing.*

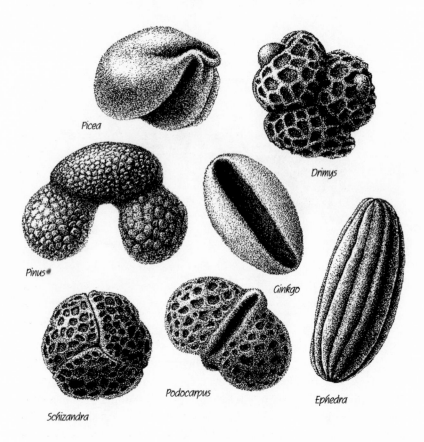

Picea

Drimys

Pinus

Ginkgo

Schizandra

Podocarpus

Ephedra

plants still invest in the "all sex and no show" strategy of producing masses of low-cost gametes instead of high-priced, showy sexual attractants for the few. The great circumboreal pine, fir, and spruce forests of Canada and Russia are living examples that this saturation broadcasting of trillions and trillions of pollen grains has been a highly successful reproductive strategy for millions of years.

Of course, almost all of these pollen grains miss their mark, forming yellow bathtub rings around alpine lakes or causing inflammation in our runny eyes and noses. Although billions of pollen grains are produced per acre of pine trees each year, they may result in only tens of thousands of pine nuts, and only a few new pine seedlings. Seen from this perspective, the pine nuts in your local health food store are not only the result of a somewhat unpredictable event, but they are also very much underpriced!

Wind pollination, however, is not the complete shot-in-the-dark it was once made out to be. It is now recognized that many modern plants, both flowering angiosperms and nonflowering gymnosperms, still practice the ancient art of directed wind pollination: *anemophily*. Don't assume these wind-lovers are inept at dispersal simply because they have not yet enticed animals to do the work of transporting pollen or spores. In fact, Steve's recent collaborations with Karl Niklas of Cornell University have shown just how clever some wind-pollinated plant structures can be.

Until the 1980s, botanists had not thought much about the significance of female pine bough and cone shape in relation to pollen capture. That's when Niklas—a card-carrying paleobotanist and mathematical pioneer in the fields of aerodynamics and plant biomechanics—decided to tackle this equation-rich issue in his laboratory at Ithaca, New York. Using his knowledge of fossils, Niklas constructed scale-model clay replicas of pine cones and other reproductive organs of now-extinct genera. He was particularly interested in those cones with upright spiky projections whose functions were unknown. Reasoning that these upright filaments might work like snow fences—breaking the velocity of pollen grains so that they drifted up around the cones—he set about to test this notion empirically in his homemade wind tunnel. The clay models were fired, painted black, and set within the target region of the wind tunnel. He then gathered pollen grains from pines and other conifers to introduce upstream from the targets.

As he set the pollen into motion, Niklas created a photographic record of them by flashing a powerful stroboscopic light at intervals of 400 times per second. When developed, the negatives revealed brilliant white dots—the pollen grains—at discrete intervals, a choreographic chart of the trajectories of individual grains as they danced their way through the complex airflow patterns created by the targets. On videotape, some of the pollen grains were seen to do lovely pirouettes, slowly turning in gyres and eddies formed in the wake of the cones. While

*Below is a cluster of male flowers on a jojoba branch* (Simmondsia chinensis) *shedding their pollen grains in the wind. Opposite is a receptive solitary female flower adapted for wind pollination. The lines depict the complex airflow patterns created by the plant's flowers and leaves.*

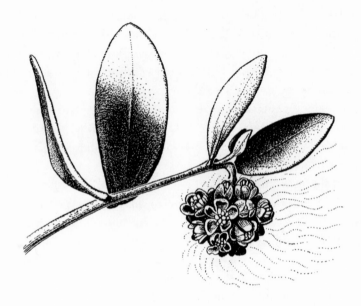

some pollen grains zoomed on, away from the clay models, others were caught in beautiful vortex patterns, finally striking, often much later, the blossom-like cones.

Reading Karl Niklas's early papers, Steve became curious about whether the architecture of wind-pollinated angiosperms could also direct the airflow of pollen around their flowers. He invited Niklas to collaborate, adapting his novel methods to study plants that Steve had worked with for years. They chose to study the famous "shampoo-bearing" jojoba shrub, *Simmondsia chinensis,* a native of the Sonoran Desert in Arizona, California, and adjacent Mexico. Above all, they wanted to know if female jojoba flowers were very efficient at capturing airborne pollen grains released from nearby "male" plants—for the jojoba is *dioecious,* meaning that different plants bear the male and female flowers.

After collecting fresh branches in Tucson and flying them cross-country to Ithaca, branches thick with pendant female blossoms were placed as the targets in Niklas's wind tunnel. The jojoba's female flowers are characteristic of many wind-pollinated angiosperms. They are small, pale green, have no showy petals, and produce no nectar, fragrance, or other attractants that might entice animals to come hither and serve as go-betweens. Each drooping flower with its trilobed stigma/style sits beneath two stiff, upward-facing leaves that look very much like the ears of a rabbit. These seemingly insignificant architectural features do much to give jojoba its strong track record among the windswept world of pollen competitors.

As the two men released thousands of jojoba pollen grains into the strobe-illuminated wind tunnel, they encountered some very unusual aerodynamic patterns before their eyes. The bunny-eared leaves above each female flower dramatically affected pollen capture. The sheetlike airflow around the leaves creates a sharp downward shift in pollen trajectories. As the air swirls around the bunny ears, its speed is broken, resulting in a veritable pollen shower onto the receptive linear stigmas and effective fertilization—so effective, in fact, that jojoba may have abandoned the insect-attracting strategies of its ancestors to rely solely on its ergonomic architecture and the desert breezes.

Although wind pollination does offer success for a relatively low energy investment per flower, pollination by animals will work in places where the wind will not. The primary competitive disadvantage for wind-pollinated conifers is that they must grow in fairly dense stands if their "pollen rain" is to be effective. Consider a solitary pine tree growing in isolation as a result of earlier long-distance seed transport by a Clark's nutcracker or pinyon jay. It may develop into a fine upstanding sapling, but it has little chance of directing pollen from elsewhere toward its female cones, nor of winning the lottery with its own pollen dispersal. For jojoba, additional problems may arise. If too many females establish themselves in the same patch where only one or two males exist, there may not be enough pollen to go around. If too many males clump together, they may swamp a lone female in too much pollen relative to the number of seeds that a sole female can produce.

In contrast, traplining hummingbirds and bees on a regular feeding route have the wherewithal to custom-deliver pollen to the door, even when considerable distances are involved. Our friend Josh Kohn has found solitary squash bees in Arizona's Chiricahua Mountains moving pollen from one wild gourd patch to another, over a mile and a half away, when female flowers there had no local pollen available to them. Such directed transfer of pollen allows outcrossing plants to be more widely dispersed. Hypothetically, a forest or desert can have more animal-pollinated species coexisting together in one habitat than wind pollination alone would allow.

When Field Museum botanist William Burger was asked why there are so many species of flowering plants, he imagined the following evolutionary scenario in which increasingly specialized species compete for floral resources:

> Insect pollinators could search and find isolated flowering individuals and thus maintain gene flow where wind-pollinated plants could not. . . . Al-

though [many] factors played a role in the achievement of angiosperm dominance, the most important innovation may have been pollination by insects, which allows outcrossing sexual reproduction in highly dispersed populations of relatively few individuals.... [There] a high premium can be placed on any factors allowing a lineage to remain effectively reproductive at low population levels.

In short, animal pollinators allowed many kinds of flowering plants to coexist, instead of letting a few densely populated species dominate vegetative cover by outcompeting other plants. A controversy regarding the advent of this innovation still rages, however, galvanized by a recent discovery in a petrified forest.

A fossil recently unearthed in Arizona has pushed back the probable date for the origin of bees and wasps, causing both entomologists and botanists to scratch their heads over the origin of flowering plants and their sexual go-betweens. Over the last few years, Tim Demko has found more than 100 fossilized bee nests inside giant silicified logs at the Petrified Forest National Park in northeastern Arizona. These nests have been enough to throw a monkey wrench into the theory of bee origins that has prevailed over the last 5 decades. Demko encountered branched nest burrows, probably made by early bees, as well as a scatter of cocoons constructed by wasps, all preserved within immense fossil trunks of a tree tentatively identified as a kin to *Araucarioxylon*. What is so amazing about these new finds from the Petrified Forest is their extreme age. The fossil trunks have been dated from the Triassic between 207 and 220 million years before the present.

That fact has little chance of making you giddy unless you know that fossils of adult bees of any age—or for that matter, fossils of their larvae and burrows—are extremely rare in the geological record. Prior to Demko's discovery, the oldest undisputed bee fossils were 80 million year old specimens enshrined in golden tombs of amber—the fossilized resins and saps of certain trees—from beach and inland deposits of Cretaceous age in New Jersey. The ancient bees from the New Jersey seaboard were highly social and belong to the genus *Trigona,* the stingless honeymakers kept today by Mayan Indians in southern Mexico. The fossils unearthed by Tim Demko and his colleague, Stephen Hasiotis, are much, much older. If further studies confirm that they were indeed made by ancient bees (no fossil bees have been found in the tunnels), they predate all other such fossils and force us to rethink the evolution of bees and flowers.

The oldest wasp body fossils date to about 116 million years ago, the oldest undisputed bee fossils to 80 million, and the oldest wasp nest at about 75 million. The angiosperms, or flowering plants, are known from fossils 110 to 120 million years old, although they may have evolved somewhat earlier. Their origin was called an "abominable mystery" by Charles Darwin in the nineteenth century. Now we are faced with the possibility that bees may have been buzzing around 140 million years before *that*. Is it possible that bees evolved so much before flowers appeared and brightened up those ancient landscapes?

That very idea was once unthinkable. In fact, it still raises the hackles on many entomologists and botanists. That is so because most biologists still tacitly believe that bees were necessary to spur the sudden adaptive radiation of flowers, an explosion we can observe in the Cretaceous sediments. Bees are thought to have played a major role in this early diversification of the flowering plants. Accordingly, either the flowering plants evolved much earlier than anyone has guessed or the earliest bees did just fine without them. They may have dined upon pollen from early gymnosperms such as ferns, cycads, conifers, and related lifeforms before true flowering plants had begun to display their eye-catching colors and food rewards. Demko and Hasiotis boldly suggest that the showy flowering plants may even have evolved to lure bees, as pollen vectors, away from other earlier seed plants.

Nevertheless, the most ancient *living* lineages of angiosperms are pollinated by beetles, thrips, primitive "biting" moths, and flies. Few of these plants have foolproof means of directing pollen for any distance from one plant to another of the same kind. Nevertheless, they must have become fairly dependable at moving sufficient pollen to the right place at the right time. Angiosperm fossils from the early Cretaceous show flowers having broad, leaflike stamens, clearly adapted to pollen foraging by insects. The earliest angiosperm pioneers—much like the living genera of *Degeneria, Drimys, Magnolia, Zygogynum*—have flowers that attract insect visitors with winelike fragrances and provide refuges to visiting insects seeking to escape predators. In turn, beetles and their kin evolved sensory mechanisms that made them more alert to floral lures and rewards. Beetles were not necessarily the sole pollinators of the earliest flowering plants, but they were certainly among those, including flies and others, attending the earliest "floral banquets" of pollen grains rich in starches and nucleic acids.

As insects were more regularly rewarded with more nutritious treats for their pollen dispersal services, some of them began to specialize by seeking particular kinds of blossoms. And this likely led to a greater diversity of floral shapes, sizes, and blooming times among flowering plants. In the 1970s, a paleoecologist named Hughes charted the increase in flowering plant species through time. Based on fossil evidence, he recognized only 500 distinctive kinds of plants in place by the late Carboniferous (286 million years ago). During that time, wind pollination was still the norm, but beetles had already begun to collect the protein-rich pollen of seed ferns, slopping a little of it around from one plant to the next.

By the early Cretaceous—roughly 144 million years ago—some 3,000 kinds of plants were present, and a portion of them were regularly hosting visits by beetles, flies, butterflies, moths, ants, and early wasps and bees. Insect pollination of plants probably began earlier, though, and was not initiated at the same time that flowering plants began to diversify. Nevertheless, some groups such as butterflies did diversify in synchrony with plant diversification, first feeding on the widening variety of nectars, then becoming regular pollinators of the same plants. The evolution of complex flowers with rings of petals—or with petals fused into tubes—was likely spurred by the variety of insect visitors active in the late Cretaceous, less than 100 million years ago. By the end of the Cretaceous 65 million years ago—when nectar production by flowers had become more common and pollinators such as bees had diverged into their modern families and many genera—at least 22,000 plant species had developed.

Since the late Cretaceous, there has been an explosive diversification of angiosperms, resulting in estimates of nearly 250,000 species present on the earth's surface today. Deep, funnel-shaped flowers appeared early in the Cenozoic, and some of these forms began a coevolutionary dance with hummingbirds. Other vertebrate pollinators—bats, honey possums, orioles, honeycreepers, opossums, lemurs, and monkeys—have affected floral evolution largely within the last 50 million years.

Today, in tropical lowland rainforests of the Americas, vertebrates likely pollinate about 5 percent of the canopy trees and 20 to 25 percent of the subcanopy and understory plants. In the same forests, bees and other insects may work 95 percent of the canopy trees and 75 percent of their shade-tolerant underlings. In attempting to gather such percentages for a variety of forest sites in the neotropics over the last two decades, Kamal Bawa of the University of Massachusetts has determined that less than 3 percent of all tropical lowland forest plants rely on the wind for pollination. In places such as La Selva Biological Reserve in Costa Rica, the wind has been documented as the pollen-dispersing agent for only 7 of the 507 plant species, not even 1.5 percent of the flora. Clearly, insect pollinators reign supreme in the most diverse tropical forests: bees, hawkmoths, beetles, moths, butterflies, and wasps each pollinate at least as many species as the wind does.

Pollination ecologists over the last century have collectively outlined certain trends in plant/pollinator coevolution that have dominated the last 135 million years. First, the angiosperm flower as we know it today evolved during the Cretaceous—at a time when insect pollination was on the rise—suggesting that many of its basic features may have co-evolved with insects. Second, these flowers began to rely on insect partners, *mutualists,* for a variety of services they couldn't get on their own. Finally, to attract the attention of these mutualists, the flower has generally had to offer some kind of reward—something the mutualists can't find on their own.

But whatever the animal pollinator—invertebrate or vertebrate—these floral rewards for coevolved mutualists cost something that wind pollination does not. An animal-pollinated plant may save some energy by having to make less pollen per flower, but it must invest even more of its resources in modifying its floral traits to attract and reward certain pollinators. Some plants invest a considerable portion of their annual energy budget in producing flowers of the appropriate color, size, shape, and odor to attract pollinators. Once the animal arrives, it must be able to reach nectaries that are pooling nutritious nectar at the right time, and then it must brush against or buzz anthers that are releasing pollen as well.

The high costs of advertising and rewards associated with successful animal pollination and fruit production take their toll on the growth of some plants. Boston University's Richard Primack and Pamela Hall have tracked the growth and reproductive success of the pink lady slipper orchid, *Cypripedium acaule,* for several years running in the hardwood forests of eastern Massachusetts. This rosette-forming plant produces a single, nectarless, pink flower on a foot-tall stalk—the flower stalk, petals, and reproductive organs amount to 18 percent of the plant's dry weight. Primack and Hall estimate that if an average-sized plant successfully attracts bumblebees, is pollinated, and produces a single capsule full of a few thousand tiny seeds, this effort will reduce its leaf area by 10 to 13 percent the following year—and reduce its probability of flowering next season by 5 to 16 percent. One successful flowering and fruiting episode can tax the entire plant for up to four subsequent years.

And yet, for the pink lady slipper orchid, there may be good reason to invest that much energy in a showy flower that results in but one fruit per year—and only seven fruits for a clump of 64 plants over the entire four-year period. At least at the present time, there is an apparent scarcity of bumblebees in the eastern Massachusetts forests, perhaps due to past pesticide spraying or other human activity. Only 2 percent of the pink lady slippers are visited by bumblebees in a way that results in pollination and fruit set. If such a showy flower cannot attract bumblebees to its pollinia, how are less extravagant flowers faring?

Martin Burd—a University of Wisconsin ecologist—is amazed that this enormous investment in lures and rewards still sometimes fails to attract sufficient numbers of pollinators. Burd has decided that this fact alone tells us something very significant about plant/animal interactions: "The very showiness of flowers might suggest that pollinator service is not obtained easily." And that fact—that pollinators are often in limited supply compared to their demand—is one that is making a number of ecologists quite uneasy. Why? Because the naturally low numbers of certain pollinators are being further diminished by man-made forces working on both agricultural and wild landscapes on almost every continent.

*Living jewels of the neotropical rainforests, a cluster of flying "orchid bees" (Euglossa spp.) hover and alight on the flowers of an orchid (Clowesia thylaciochila). The enlarged inset shows a bee that has just visited a flower and bears the badge-like yellow orchid pollinia—sacs containing the pollen grains—on its thorax.*

# CHAPTER 3

# *Pollinators*

*Waiting for the Bait to Pervade the Air*

*S*TEVE REMEMBERS:

It was early in the rainy season, when bouncing over Panama's always muddy Pipeline Road was still fairly easy—we might even arrive in time to do our work as planned. My friend behind the wheel was accustomed to working conditions far worse than these. I was intimate with this narrow winding road, too, having many times ridden shotgun as the green Toyota's massive knobby tires followed the ruts cut through the last remaining primary tropical lowland rainforest in the Panama Canal Zone—the Parque Nacionál Soberania. We called the driver "Mad Dog" Dave Roubik, for he was one of the gutsiest tropical biologists there was. In scholarly circles, Dr. Roubik was well known for his research on the so-called stingless bees, which pollinate many of the plants in this forest. Roubik's eclectic interests extend far beyond the stingless, however, and range from the infamous (stinging) Africanized bees to modern jazz. This day, I was going to help Dave Roubik put out scented baits to entice the males of a flashy group of bees known as *euglossines,* the flying jewels of the rainforest.

The rain was already falling by the time we got into the Toyota before dawn. Its patter—augmented by the din of cicadas and katydids—nearly drowned out the jazz riffs blasting from Roubik's stereo as we wound toward our makeshift work station at the Kilometer Ten roadside marker. Suddenly Dave braked furiously, throwing me against the glove compartment where he kept his stinky baits. I looked up in time to see the giant spotted cat we had nearly run down. My heart was pounding as I realized that the cat disappearing into the undergrowth was a jaguar—the first that either Dave or I had seen in the wild. It was one more indication that the Soberania National park has retained patches of virgin tropical forest, a rarity in Panama today.

At last we parked, and began the task at hand: taking the monthly orchid bee census, as Roubik and colleagues have done for an unbroken string of over 100 straight months. In this densely forested habitat, we would not have to chase down the bees at flowers; they would seek us out. We would attract them with three synthetic chemical baits: cineole—with the medicinal smell of Vicks Vapo-Rub; methyl salicylate—with a strong candy aroma, like wintergreen mints; and the infamous skatole—with the penetrating odor of fresh, warm, mammalian dung.

Roubik kept all three scents in Tupperware containers, but took special care with skatole—its container screamed to the curious, "STENCH—KEEP TIGHTLY CLOSED!" One little slip and this pungent aroma would stick to our clothes and skin for days. But Roubik had run through this entomological routine so many times that it was now second nature to him. We offered up the three baits on blotter papers tacked to the roadside trees at breast height. And then Dave and I unfolded our aluminum patio chairs in the middle of the road, to watch the flight of electric-blue morpho butterflies, to listen to the thrumping drone of cicadas, and to wait for the metallic-hued male orchid bees to follow the airborne scent trails to our baited blotters.

We did not have to wait very long. Within the first five minutes I heard the high-pitched wingbeat of an inquisitive male—an all-green *Euglossa imperialis,* dangling a proboscis as long as its body. Soon other male orchid bees arrived at the blotters, shrouding them with their brilliant metallic greens, purples, coppers, and golds, flashing like gemstones in the lurid tropical sunlight. Most would alight briefly, though some would wander around on the blotters for several minutes if left undisturbed.

Occasionally, Dave would get up from his chair, walk over to one of the blotters, then quickly grab an orchid bee between his fingertips in order to identify it. After handling and releasing tens of thousands of these bees over the years, he could quickly identify any of the 57 species that lived nearby in this rainforest. All of them use specialized hairs on their front legs—flattened into scrapers—to rub the scent out of the tattered blotter fibers. They were eagerly collecting the volatile scents on the blotters, for they recognized them as the same chemicals present in the majority of orchids in the Soberania rainforest.

The message was clear: these scents were signposts for the bees the same way the Kilometer Ten marker directed Dave and me. It remains unclear, however, how the bees actually *use* the chemicals they collect. What advantages could the bees possibly gain by adding certain scents to their sexual behavioral repertoire?

Euglossine bees were relatively unknown until the late 1960s, when Calloway Dodson and Robert Dressler used gas chromatography to analyze the chemistry of certain spicy fragrances from neotropical orchids—the key, they believed, to euglossine ecology. Until that time, there were few specimens of male orchid bees in museum insect collections, for no one had discovered a foolproof way to capture them. (This remains especially true for females to this day.) After elucidating some of these pungent chemical brews of mint, cinnamon, and vanilla-like scents, the scientists noticed that male orchid bees would congregate in great numbers near any material doused with these simulated floral fragrances. They began tacking up blotters—each drenched in a particular chemical—to the bark of tropical trees, as Roubik has done in Panama and as we have now done at the northern limits of tropical forests, where they meet the deserts of Sonora, Mexico. *Et voila!* Little-known bees seem to come out of the woodwork, adding to the diversity of euglossines unimaginable before the scents became widely deployed by curious naturalists.

In the early 1960s, scientists at first believed that males were lured to orchids lacking food rewards because the flowers mimicked the resinous nests constructed by female bees. Later the floral fragrances were proposed as attractants, since it appears that male bees use the fragrances as a focal point around which they might congregate, establishing a rendezvous site similar to the *leks* used as display sites by birds such as sage grouse. As more information accumulated, however, it became clear that these bees have come to depend on the spicy perfumes of orchid flowers as precursors for their own sex pheromones to lure females to them for mating. In a strange twist of nature, the bees may be stealing the orchid's sexual message and transforming it into their own aphrodisiac.

In order to *utilize* the fragrances scraped up from orchid petals (or blotters), the bees somehow "process" the volatiles in spongy, enlarged glands in their hindlegs. Apparently the bees mix them with their own chemical secretions or otherwise transform them biochemically. In any case, when presented in aerosol form by male bees congregating and visually displaying at a lek, such aphrodisiacs are irresistible to their potential mates.

Parts of this story remain conjectural. It is no easy task to confirm under rainforest conditions that female euglossines choose their mates on the basis of "who's wearing the best aftershave." Such mate selection by scent, however, would not be all that surprising, since bees are chemists par excellence. In fact, there are other well-established examples of olfactory mate choice among bees. Female carpenter bees *(Xylocopa varipuncta)* in the Southwest choose their mates after being attracted to their sweet perfumes from far downwind. This has been demonstrated, using a synthetic male carpenter bee pheromone to attract females of this species, by Steve and his field associates—William Wcislo, now with the Smithsonian Tropical Research Institute in the Republic of Panama, and Robert Minckley at Auburn University—for several sites in the Sonoran Desert near Phoenix and Tucson.

The very density of bees at each blotter-paper bait makes it easy to see if the euglossine males have been actively pollinating nearby plants. Of course, these temporary gatherings serve only as indirect signs of their local abundance on the flowers themselves. To do better, you would have to climb high into the rainforest canopy and look for blooming orchids with signs of pollen removal. There are few easy ways to obtain direct evidence of bee visitation to these *epiphytic* flowers—flowers that root or suspend themselves from high up on the trunks of trees.

Once a bee is captured, you can easily see large yellow paired protuberances jutting out from its head. Each one consists of a fresh *pollinium*—the bright yellow pollen sac that attached to this male while it was mining an orchid for its delicious sexual perfumes. Orchid pollinia are marvelous inventions, a special way of packaging pollen grains in shrink-wrapped containers for transport by animals. Orchid pollinia look like two bright yellow egg yolks attached to a leathery but pliant brown stalk. As a euglossine bee forces its way into an unvisited orchid flower, pollinia get stuck to its armor, often in hard-to-reach "safe sites" where the bees cannot remove them even with the most vigorous foreleg action—just as you can hardly scratch an itch in the small of your back. For days on end, an orchid bee may have pollinia stuck to him in places that may seem comical to humans—between his eyes, atop his head, protruding from his abdomen—before he visits another orchid flower of the same species, which can detach the reward from the hapless go-between. When a euglossine male crawls into another orchid flower on its back, abdomen, or side, depending on the flower shape, the pollinia literally become "unglued." Pressure from the flower forces the two pollen sacs to be left behind, rubbed into a receptive, gooey stigma, effectively ensuring cross-pollination.

In the years that Roubik and colleagues have inspected bees month after month, they estimate that much less than 5 percent of the bees visiting the blotters have recently visited an orchid—their pollen sac "badges" and glue marks are missing. The scientists wonder why so few individuals are carrying orchid pollinia around. This impression is generally consistent with other orchid pollination studies: not all bee visits result in successful pollen removal or transfer from one flower to another. Even though Roubik's studies indicate that Soberania's orchid bees have more stable year-to-year populations than almost any other insect ever studied, there is no indication that bee visits and availability of orchid pollinia are in some kind of "perfect balance" as implied in early textbook treatments of mutualism between plants and animals.

Although euglossine bees depend on pollen and nectar harvested from other plant families for their nutrition, they have many specialized adaptations that help them benefit from orchid flower fragrances. In turn, tropical orchid floral shapes, pollen sac size and placement on bees, as well as pollinial "glue," all indicate some level of the flower's adaptation to euglossines as pollen dispersal agents. Nevertheless, the "just-so" accounts of each euglossine bee fitting into a one-to-one partnership with a special orchid host have not held up to recent scrutiny by floral ecologists.

We now know that a single species of orchid is often visited by a "guild" of euglossine bees, often from several genera, each of them attracted by particular scents. A bee may visit several species of orchids, fungi, or even oozing tree wounds to gain its own peculiar mixture of scents. In turn, the orchids have developed ways of ensuring that their pollinia get transferred *only* to another flower of their own species and are not "wasted" by being left in flowers of an altogether different species. Every orchid genus—and sometimes certain species as well—has a predetermined "map" of locations on the body of its bee visitors on which to glue its precious pollen cargo. During each subsequent encounter with a flower of the same kind, the particular shape of the flower forces the bee to twist and turn in a way that increases the probability that the properly placed pollen package is effectively transferred. This not only guarantees pollination and subsequent fertilization; it also guarantees outcrossing between plants of the same species. No matter how many bee species visit the same orchid, they all wear its pollen sac badges in the same place, before surrendering them to the orchid's kin.

For flowers, the benefits gained from animals are clear: the transport and appropriate placement of pollen—that's why we call these animals "pollinators" and not simply "floral visitors." But what do animals gain from flowers? This is a much stickier question. Its answers are as varied as the benefits plants provide: shelter from storms, protection from predators, safe refuges for mating, strategic stakeouts for territorial defense, spots for ambushing prey, as well as delectable sources of nutritious nectar and pollen, resins, oils, drugs, perfumes, and other novel chemicals. And, as we have seen, some euglossine bees may transform orchid fragrances to make their own sexual lures.

If these are the benefits, what are the costs? Biologists sometimes define the "cost of pollination services" only in terms of the cost of calo-ries burned by pollinators moving from one flower to the next within their territory—along their "trapline." For example, the travel costs of hermit hummingbirds moving along a trapline between *Heliconia* "bird-of-paradise" flowers may be high, for these plants are often more than a hundred yards apart, taking the hermit through nearly as much as a mile of forest flyways to obtain enough nectar to fuel its morning journey.

But the varied costs of pollination services can also be considered in evolutionary terms—that is, as the investment in specialized morpho-logical and chemical adaptations and the limits that specialization places on making use of *other* resources. To obtain the nectar of one particular *Heliconia* species, for example, its attendant hummingbirds must rotate their heads at an angle of more than 90 degrees—almost upside-down—in order to allow their bills to reach the hidden nectaries. Of course, their bills must be of the appropriate length and curvature as well. The *Heli-conia* offers copious quantities of energy-rich nectar to hummers who can achieve this acrobatic feat. As ornithologist Steve Hilty has noted, "In doing so, pollen is placed on the chin and base of the bill, thereby reducing the possibility that [this *Heliconia's*] pollen will be deposited later on the wrong kind of flower." In other words, as the pollen-carrying humming-bird probes the right kind of floral tube, the flower's shape dictates the precise transfer of the pollen from the chin or bill; in the wrong kind of flower, the pollen may be rubbed off, but it will be deposited in the wrong place, far from the receptive stigma.

For pollinators, therefore, the costs of visiting flowers can be thought of in many different ways. In addition to the immediate en-ergy costs of flying from one flower to the next, there are also the sub-sequent behavioral and metabolic investments in gathering, trans-porting, and processing floral rewards. The orchid bees must not only carry pollinia; they must also process floral fragrances before they have adaptive value. Their time-consuming visits to orchid flowers also make them easy targets for predators—ambush bugs and crab spi-ders—or parasites such as mites. These costs are all fairly observable: a biological accountant can roughly calculate the amount of time a eu-glossine bee spends foraging or the probability that it will be attacked by a predator or parasite.

But the *evolutionary* costs of focusing on orchid rewards are difficult for even the most capable accountant to project. Euglossine bees are high-energy flyers, capable of running long traplines, but they are also equipped with sensory apparatus to spot orchids along the way. They have specialized organs—proboscides that look like fishing poles—allowing them to reach foodstuffs hidden deep within flowers. It is likely, too, that much of the bee's orchid-fragrance processing and use as an aphrodisiac is genetically programmed. Presumably, by specializing early in their evolutionary lineage on collectable sex attractants from orchids and other sources, these highly specialized apid bees became locked into their roles as scent harvesters. And how are male orchid bees to divide their time between flowers that are rich with fragrances and flowers that offer nutritional nectar? (Males aren't concerned about collecting pollen from flowers.) This is a choice in foraging time investment that few other bees must reckon with.

For a few other sets of plants and pollinators, however, the evolutionary costs of such an obligatory mutualism have been roughly calculated. This is easiest to do for insects that are rewarded for their pollination services with nutritious seeds for their young, as in the case of the Finnish buttercup, *Trollius,* and the flies that mate in its flowers. The number of fly eggs found per flower almost perfectly matches the calculated balance between costs and benefits presumed to maintain this mutualism over the long haul.

Yet the classic example of this "seed for seeds" sort of mutualism is the desert yuccas and their moth pollinators, an obligate dependency first discovered in 1876 by Charles Valentine Riley, just two years before he became national entomologist for the United States government. Building upon Riley's initial meticulous observations, scientists such as John Addicott, Judith Bronstein, and Olle Pellmyr have demonstrated how in some species, female moths exchange the placement of pollen on yucca stigmas for a small share of fertile seeds to feed their young. These female yucca moths have specialized mouthparts that allow them to collect pollen from the pale, velvety yucca flowers—pollen they then carry around with them for most of their brief lives. There are "mimic" or "robber" moths that take advantage of these mutualisms, too, but they have not yet been found in all yucca species.

*An example of mutualism occurs in the Chihuahuan deserts of Arizona, New Mexico, and Mexico where yucca plants like this* Yucca elata *are pollinated by specialized moths of the genus* Tegiticula. *Moths actively collect pollen grains, grasp the ball-like mass with their elongate mouthparts, and force it into a receptive stigma. We are aware of no other pollinator that intentionally pollinates a flower. This moth does so to ensure a supply of seeds for its voracious larvae developing within the immature fruits.*

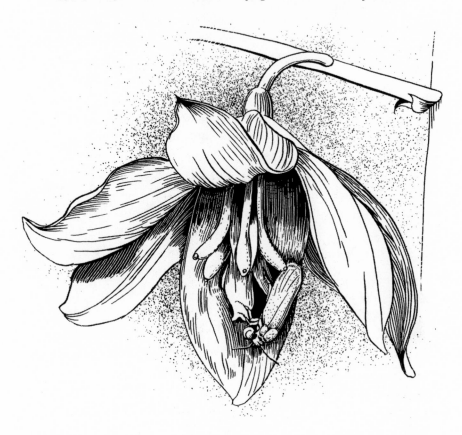

Vanderbilt University's Olle Pellmyr has described what happens after the moth cuts into the yucca flower ovary and deposits her eggs within:

> She then walks up to the stigma, and actively deposits a small amount of pollen. In so doing, she insures that lack of pollen will not limit the availability of developing seeds, which is the exclusive food of her progeny. . . .

Upon hatching, the larva consumes only a small portion of the developing seeds, leaving many intact. Reciprocal plant specialization through exclusion of other pollinators has led to obligate mutual dependence between yuccas and their pollinators, as neither species can successfully reproduce without the other.

According to Pellmyr, that's where the calculation of energetic costs gets very interesting. If the moths could increase their egg loads per fruit, they would benefit, but the plant's reproduction would likely suffer from fewer germinable seeds resulting from the moth's greater success. Instead, Pellmyr and his colleague Chad Huth have found that yuccas tend to shed flowers that have high numbers of eggs planted in them, especially if the flower has received few *pollinations* by the moths. In other words: the moths that produce just enough eggs to get by, but repeatedly and dependably pollinate the same flower, are favored by natural selection. Evolutionarily, moths that overburden a developing fruit with too many eggs, and are slackers when it comes to providing pollination services, are disadvantaged: the flower aborts, the seeds fail to develop, the moth larvae are left without food, and they die. This selective fruit abortion keeps the symbiosis between moths and yuccas stable over evolutionary time, so that the fitness gains of one species do not occur at the expense of the other.

Of course, the ultimate cost of this mutualism is the most obvious one: each of the mutualists suffers reproductive failure when the other can not be found at the right place at the right time. Judith Bronstein of the University of Arizona has observed such an occurrence on the grassland edge of the Santa Rita Mountains, just 30 miles south of her lab, where she regularly works on the biology of the Spanish bayonet, *Yucca elata*. Bronstein records: "The yuccas in the Santa Ritas bloomed a month late last year, and ended up out of synchrony with the emergence of most yucca moths. As a result, the yucca fruits which set seed were the very earliest ones; late-blooming plants failed utterly."

We have observed a more permanent disruption to the yucca/moth mutualism while working in Papago Park, Arizona, a degraded natural area now in the middle of metropolitan Phoenix. Up until the 1930s, photos of Papago Park show that banana yuccas, *Yucca baccata,* were a prominent feature of the natural vegetation there. Perhaps because of an explosion in rodents and rabbits due to the depletion of predators in the area, it appears that yuccas were simply browsed out of existence. The only yuccas remaining in Papago Park are those planted in the desert botanical garden there in the 1940s and 1950s, and these bloom in patches, and not every year. The cultivated yuccas in the garden *never* set seed—either the moths locally died out as the natural population declined or the current bunch of blooms is not large enough to sustain a viable population of moth larvae.

One of the most ironic evolutionary costs of being a pollinating partner in a mutualism is that you are vulnerable to "cheaters" and "robbers" such as the additional moth species now being found associated with the yucca/moth partnership. These opportunists usurp the benefits of the floral host without providing reciprocal benefits. There are robbers and cheaters, too, that take advantage of bee, moth, and hummingbird mutualisms with various flowers. For instance, certain saw-billed hummers cut into the corollas of tropical flowers to obtain nectar while bypassing the pollen altogether—thereby depleting the food available for the euglossines or hummers that approach the flower from its opening.

Not only can competition from cheaters and robbers limit the resources available to a faithful mutualist pollinator, but it can reduce the fruiting success of the flower to which it has provided pollination services for millennia. Fortunately for the plants, the numbers of cheaters and robbers seldom overwhelm the number of mutualists drawing on the same set of flowers—otherwise, the coevolved community might collapse. Undoubtedly there is a threshold density of host plants, a point below which such "parasites" cannot be supported in their game of deceit. Mexican ecologist Francisco Ornelas has determined that many nectar-robbing hummingbirds are far less abundant than those species that provide pollination services:

> One trade-off for being opportunistic—obtaining nectar by bypassing the opening used by pollinators—is *rarity*. Individuals of a third species take advantage of mutualistic partnerships and become density-dependent on the mutualistic interaction. Rarity among nectar robbers would be the

evolutionary outcome of plant specialization on a particular set of true pollinators.... Nectar-robbing hummingbirds seem to be rare in nature and/or
erratic on a regional scale.

Any disruption between the plant and its true pollinators that reduces floral abundance will endanger the robbers and cheaters also.
The natural rarity of nectar-robbers can therefore be aggravated either
by deforestation that directly destroys its nectar sources or by a decline
in the legitimate pollinators that ultimately leads to reproductive decline in the host plant population.

Obligate mutualisms—in particular, one-to-one symbioses between
a single pollinator species and a single plant species—make up only a
small percentage of all plant/pollinator interactions: the risks of overdependence must be too high for many species to take. Nevertheless, even
though obligatory partnerships are rare compared to interactions
involving generalists, they often have a disproportionately large influence in structuring plant and animal communities. The best example of
this is found in the world's wet tropical lowlands, where strangler fig
trees play a key role in forest spacing and the support of fruit-feeders.

In countries ranging from Peru to New Guinea, figs are a critical resource for a wide variety of animals in tropical forest communities—
from bats and primates (including people) to parrots and birds of paradise. Up to 70 percent of vertebrate diets in certain forests is derived
from figs. There are more than 750 different fig species in the world,
and the majority of them rely on different species of tiny wasps as their
exclusive pollinator. And, in turn, these fig wasps depend on a portion
of a fig's developing seeds as their food at a critical stage in their lives. As
Judith Bronstein observes:

> Figs have been called keystone mutualists of tropical forests—a keystone is
> the one that holds the arch together. The idea here is that if you pulled out
> the keystones, it would be disastrous for many of the animals that rely on
> them.... This could happen, for instance, due to the selective logging of the
> trees upon which strangler figs establish themselves. Or, by spraying insec-
> ticides, you wipe out wasps, which in time will cause their fig tree hosts to
> decline or possibly go extinct since the trees won't be able to rely upon any
> other local pollinators. They must rely upon the very highly specialized fig

*One of the best examples of reciprocal coevolution is found in a fig inflorescence (*Ficus carica*) and the minuscule fig wasps (*Blastophaga spp.*) that pollinate these largely tropical rainforest trees. Here we see the complex fig anatomy with its hundreds of flowers, a wasp laying an egg inside a flower, a wingless male wasp, and a new generation of fig wasps emerging through the wall of an older fruit.*

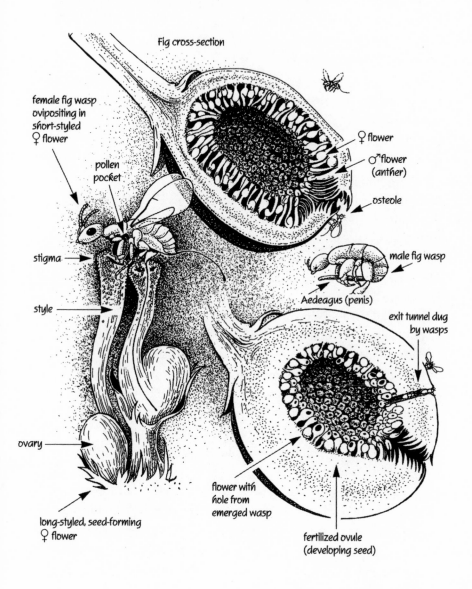

Fig cross-section

female fig wasp ovipositing in short-styled ♀ flower

pollen pocket

♀ flower

♂ flower (anther)

osteole

stigma

style

male fig wasp

Aedeagus (penis)

exit tunnel dug by wasps

ovary

flower with hole from emerged wasp

long-styled, seed-forming ♀ flower

fertilized ovule (developing seed)

wasps for their reproductive success. Both mutualists are locked in an intricate evolutionary dance and cannot change partners. . . . It's assumed that populations would crash if figs or their obligate mutualist pollinators were wiped out. If this happened, cascading extinctions would be expected. One would assume that obligate parasites of the mutualism would go extinct, and monkeys would radically shift their diets or starve. With fewer animals feeding on fig fruits and seeds, perhaps there would be reduced food for predators such as raptors and jaguars. This isn't a trivial example but a very real threat due to tropical deforestation.

In attempting to predict the effects of rainforest fragmentation on figs and their associates, Bronstein and her colleagues have asked: "How many fig trees are necessary to allow a wasp population to persist?" Initially they estimated that between 95 and 294 trees of one particular fig species would be required to sustain a pollinator population for a minimum of four years. Bronstein's colleague Doyle McKey at Montpelier later estimated that a minimum of 300 adult fig trees would be required to ensure a 99 percent probability of fig wasp survival. McKey then asked: "What area of forest must be preserved to ensure the maintenance of minimum viable populations of figs and wasps?" His answer was disconcerting. Depending on the species, 800 acres to 800 square miles might be required to ensure the long-term survival of fig, wasp, and all the other dependent birds and mammals.

More often than not, the risks associated with such tight dependencies are too high to encourage such strict obligate mutualisms. Instead, we find that the majority of mutualists can be considered products of *diffuse coevolution*—that is, a *guild* of pollinators collectively focuses on a *set* of particular flowers that have similar floral forms and presentations but may either bloom sequentially or overlap in space and time. If either the flower or pollinator does not show to play the mutualism game, both have other options.

Of course, not all pollinators necessarily specialize on a particular set of like-shaped flowers: some, like the honeybee, are jack-of-all-trades generalists. To see just how eclectic a pollinator may be, let us shift from the rainforest back to the Sonoran Desert where we live and work, and consider a study of honeybee responses to a diversity of floral pollen rewards over the last decade.

*S*TEVE REMEMBERS:

*Ouch!* I paused to disentangle my shirt sleeve and a bit of flesh from a catclaw acacia branch as I traversed the rocky bajada slope above Pima Canyon. Here, in the Santa Catalina Mountains not more than 15 miles away from Saguaro National Park and downtown Tucson, I have been sampling five managed European honeybee colonies weekly for the past ten years to check the rich pollen harvest by local honeybees. Although it was a cool morning in mid-November, the honeybees were flying around like great streams of miniature aircraft. As I came upon the hives, I heard the wingbeat buzzes of thousands of airborne worker bees, active here at least 11 months of the year. There was every indication I would certainly find new pollen to sample from these hives.

Unlike honey gathering, collecting pollen samples from beehives is easy—it doesn't aggravate the bees inside the way tearing their home apart to steal away with 50 pounds of honey does. Without wearing a bee suit or a protective veil, I simply walked slowly and deliberately past the lines of streaming bees, and crouched down behind the hives. There I simultaneously pulled open the drawers of wood and hardware cloth devices known as pollen traps. In each of them were low, rounded mounds of heaped corbicular pollen pellets in a rainbow of colors. The bees had indeed been busy.

There were thousands of these pellets, all roughly the same size, but imbued with hues of yellow, orange, green, brown, and white—each kind dominated by a particular kind of pollen from a neighboring wildflower. That was curious, since I had seen few flowers on my hike up into the canyon. Nevertheless, the bees have eyes for wildflowers that lie beyond our sight—and, moreover, they may have recently foraged within a 60-square-mile area. As E. O. Wilson once put it, if bees were the size of humans, and their flight distances similarly scaled-up, a single bee colony placed smack dab in the middle of Texas would collect pollen and nectar from wildflowers from about one-half of that state. These bees knew intimately what was in bloom at any hour of day or time of the year over an astonishingly large area of Sonoran Desert near Tucson.

Each pellet color signified that one foraging bee had found and successfully returned to her nest with two small pollen payloads. Each of these pellets contained several to tens of thousands of pollen grains glued together with bee saliva or regurgitated honey. And in these traps, over a year's time, I would eventually collect pollen pellets containing 55 flowering plant species from 40 genera in 25 families—about a fourth of all the kinds of plants blooming within their reach.

In short, honeybees have the greatest pollen dietary range—*polylecty*—of any known pollinator. They are known to collect fine particles of bizarre materials other than pollen, too, including mold spores, cheese mites, flour, coal dust, and sawdust. They even obtain pollen from normally wind-pollinated and nectarless flowers such as those of grasses and ragweeds. In fact, up to 30 percent of the pollen diet in the Sonoran Desert can come from native and introduced plants that are considered to be *anemophilous*—essentially oriented toward wind pollination through their floral structure, pollen buoyancy, and design.

From the Pima Canyon pollen traps, I discovered not only how much pollen was annually harvested by honeybees—often 50 to 85 pounds per colony per year—but also what types of flowers they visited. Daisies, mesquite, creosote bush, lilies, jojoba, triangular leaf bursage, mistletoes, and mustards made up three-quarters of the honeybee pollen diet in the Santa Catalina foothills over the past decade, but those proportions might soon be changing. Down below the bajada slope where I stood, I could see earthmoving equipment ready to scrape away the native vegetation living on my study site.

It was an indicator of what was to come, on a massive scale, to this neighborhood, and to many others. The cactus forest would soon be replaced with 500 new luxury homes, landscaped with a few showy natives like saguaro and prickly pear cacti, but also with a variety of newcomers. As Easterners, or Californian and Mexican migrants, arrive in Tucson, they bring with them a rash of exotic landscape plants—Australian eucalyptus, African sumac, olive and pepper trees. While the honeybees will no doubt adapt to them, the yucca moths and solitary bees native to habitats like this will not fare so well. And what of the private lives of the local native plants: how would they be affected by the new habitat fragmentation and continued pollination by exotic bees? A turnover in local vegetation may signal for them a demise in their food resources and could inevitably lead to a local extirpation—

the loss of a population, either temporarily or permanently from a particular place. One population's extirpation does not inexorably lead to a species' extinction. But it does pose a big question: How many populations can be lost before an entire species can be considered threatened? As we shall see, there is no single answer, but there are very clear and disturbing trends.

In the montane cloudforest near San Vito, Costa Rica, a little hermit hummingbird (Phaethornis longuemareus) *is about to visit an unusual plant* (Columnea florida) *in the family Gesneriaceae. During a brief period of the year, paired reddish blotches appear near the leaf tips—visual signposts that the birds learn to associate with sweet nectar hidden in tubular flowers on the stems below.*

# CHAPTER 4

# The Perils of Matchmaking

## Pollination Syndromes and Plant/Pollinator Landscapes

*S*TEVE REMEMBERS:

I scanned the hillsides above Napa Valley as we hiked along their ridges, a troupe of us trailing behind my entomology professor, Robbin Thorp, from the Davis campus of the University of California. Compared to the lower valley slopes and bottomlands where lush vineyards overwhelmed the remnant patches of natural vegetation, the hills were sparsely vegetated. High on the ridges, where shiny green serpentine rock came to the surface, the dwarf chaparral was thinner because of the witch's brew of weird chemicals active in serpentine soils, some of which are toxic to plants.

Although many of my fellow students commented on how impoverished the plant cover was, each widely spaced shrub or tree could be seen and smelled for its uniqueness: digger pine, Sargent cypress, the localized form of manzanita. This advantage was not lost on me. I was working at learning both the flora and the fauna of the Inner Coast Range along the continent's edge. I asked Robbin to point out any plants he noticed that were unique to the area.

About that time, a strong and unusual fragrance made me stop dead in my tracks. Although miles above any winery, I recognized the fragrance as virtually the same as the bouquet from an estate-bottled Napa Valley Cabernet Sauvignon I had recently tasted and enjoyed. Looking around, I realized that the aroma could only be coming from one place—the dark maroon flowers of the large-leaved bush alongside the trail. Not just any bush; but the western spicebush, known to botanists as *Calycanthus occidentalis.*

Even as I enjoy a glass of tannin-rich wine today, I remember the flash of olfactory recognition that leapt through my head the moment I smelled the spicebush: "Beetles. It must be pollinated by sex-starved sap beetles!" Without having seen the pollinators, I had predicted—simply by cues of floral color and fragrance—at least one of the animal groups that might be transferring pollen between plants of the western spicebush in these Napa foothills.

Table 1: Pollinator Syndromes

| Syndrome | Pollinator | Anthesis [a] | Color | Odor |
|---|---|---|---|---|
| Cantharophily | Beetles | Day and night | Variable, usually dull | Strong, fruity, or aminoid |
| Sapromyophily | Carrion and dung flies | Day and night | Purple-brown or greenish | Strong, often of decaying protein |
| Myophily | Syrphids and beeflies | Day and night | Variable | Variable |
| Melittophily | Bees | Day and night or diurnal | Variable but no pure red | Present, usually sweet |
| Sphingophily | Hawkmoths | Nocturnal or crepuscular | White or pale to green | Strong, usually sweet |
| Phalaenophily | Small moths | Nocturnal or crepuscular | White or pale to green | Moderately strong, sweet |
| Psychophily | Butterflies | Day and night or diurnal | Bright red, yellow, or blue | Moderately strong, sweet |
| Ornithophily | Birds | Diurnal | Bright red | None |
| Chiropterophily | Bats | Nocturnal | Dull white or green | Strong, fermented |

*Source:* Modified from Wyatt (1983). Nectar concentrations are those reported by Pyke and Waser (1981).
*a.* "Anthesis" means time of floral opening.
*b.* "Actinomorphic" means radially symmetrical; "zygomorphic" means bilaterally symmetrical.

This guessing game is one that has both intrigued and aggravated biologists for much of this century: to what extent is it possible to deduce the animal pollinators of a certain kind of plant simply by knowing key characteristics of blossom and creature beforehand? The interacting set of plant and animal attributes that form a consistent pattern is known as a *pollinator syndrome*. (Table 1). As we shall see, this routine of pattern recognition, like any guessing game, can be great fun, but it often underestimates the complexity and variability of relationships found in nature. Some biologists today are forsaking this game of identifying "classical relationships" for another, more difficult one. Because this newer game requires more empirical observations of pollinator foraging and flowering patterns of all the plants in a particular landscape, it better approximates the diversity of plant/pollinator interactions.

This novel paradigm, recently proposed by ecologist Judith Bronstein,

| Flower shape [b] | Flower depth | Nectar guides | Reward |
|---|---|---|---|
| Actinomorphic | Flat to bowl-shaped | None | Pollen or food bodies |
| Usually actinomorphic | None, or deep if traps involved | None | None |
| Actinomorphic or zygomorphic | None to moderate | None | None or pollen or nectar |
| Actinomorphic; held horizontal or pendant | None to moderate | Present | None (41.6%) and pollen; open or concealed |
| Usually actinomorphic; upright | Deep, narrow tube or spur | None | Ample nectar (22.1%); concealed |
| Actinomorphic or zygomorphic | Moderately deep tube | None | Nectar; concealed |
| Actinomorphic or zygomorphic | Deep narrow tube or spur | Present | Nectar (22.8%); concealed |
| Actinomorphic or zygomorphic | Deep, wide tube or spur | None | Ample nectar (25.4%); concealed |
| | Brush- or bowl-shaped | None | Ample nectar (18.9%) and ample pollen; open |

allows us to identify *plant/pollinator landscapes*. These landscape patterns inform us not only of the interaction between beetle and spicebush, but all the other flowers that beetles sequentially visit and all the other animals that visit spicebushes and their neighbors. Like a basic ground rule in the pollination syndrome game, this one also assumes there are key traits of flowering plants and pollinators that do not vary independently, but cluster together in space and in time. But unlike another assumption in the pollination syndrome game, we now recognize that such traits vary both within and between species. Not every spicebush population may flower at the same time, nor attract the same beetles. Likewise, some beetles may actively pollinate spicebushes, while others may only visit them infrequently, preferring to concentrate their activities on other neighboring plants. To play either game, we must learn which floral traits matter to pollinators and which features of animals have shaped flower morphology and phenology over evolutionary time. To highlight some of these traits, let's return to the spicebush.

Looking at the maroon blossoms of the spicebush more closely, we begin to note clues that they could indeed be beetle blossoms. The flowers are about an inch and a half long, formed by fleshy, almost succulent *tepals*—a fusion of the corolla's petals and the sepals of the calyx—that protect the rewards within. They produce no nectar but are loaded with bright yellow, rather oily, pollen grains. And on the tips of each of the innermost reddish tepals, we encounter rough white "food bodies" containing what is known as tubular pollen. Such food bodies are found within many primitive flowers, supposedly to hold the gustatory interest of beetles so they do less damage to tender plant parts as they seek protein-rich pollen upon which to dine. Held within the protective flower, beyond the reach of most predators, beetles also engage in frequent copulations with others of their kind.

These floral traits are reminiscent of the primitive flowering plants of New Guinea and New Caledonia, which have been studied as "living fossils" to reconstruct the conditions under which animal pollination first evolved. Such flowers either have sweet, spicy fragrances or strong, unpleasant ones—smelling like rotting tropical

*In the hills of California's Napa valley grows the western spicebush* (Calycanthus occidentalis). *The large flowers have many primitive features including numerous parts, fleshy tepals, and white food bodies. The floral aroma is reminiscent of a blend of overripe fruit and cabernet sauvignon wine. Dozens of nitidulid and staphylinid beetles are attracted to these flowers as pollinators.*

fruit or, worse, a pair of neglected dirty sweat socks returning from summer camp. There is one tree species in Guanacaste province of Costa Rica—*Sapranthus palanga*—that students have dubbed the "thousand-dirty-sock-tree." In either case, these open-cupped blossoms offer a number of attractants (such as fragrances) and rewards (oily yellow pollen and whitish food bodies) that beetles and flies can both perceive and handle. When beetles feed on spicebush pollen, some of it sticks to their hard cuticles. Later, when the beetles move to feed on another plant, it gets passively transferred to floral parts (including sexually receptive stigmas) in other flowers.

Beetles, however, do not always make the most effective pollinators. In fact, they blissfully browse on various flowers, many of which they never pollinate. For certain species like the spicebush, they do transport pollen in their search for more white food bodies and mates. But often they drop more pollen when they are walking between plants than they deliver to another flower. Furthermore, if another spicy species forms a local attraction, they may immediately zip over to browse there without ever delivering pollen to a compatible flower of the first species.

This sort of casual pollen transfer by beetles has been termed *mess-and-soil pollination* by floral biologists because these insects tend to eat and defecate their way through such blossoms. A single kind of flower is seldom, if ever, dependent on just one kind of beetle. The roving beetles, in turn, are hardly ever attracted to just one species' scent in their neighborhood. The mutualism is loose, diffuse, if you can call this relationship a mutualism at all. And yet it has become customary for ecologists conversant with the pollination syndrome game to refer to spicebush blossoms as "beetle flowers." The gangly blooms of ocotillos are known as "hummingbird flowers," and the white trumpets of night-blooming cereus as "hawkmoth flowers," even though scientists have recorded other animals pollinating or at least visiting and feeding among these blossoms.

This matching of blossom types with guilds of similar flower-feeding animals appears to be a universal pastime among humans. Just like players of the pollination syndrome game, many indigenous cultures assume that closely related animal visitors visit particular shapes of flowers, to feed upon their honeylike nectar. Ecologists have simply refined this folklore by acknowledging the many floral resources that attract animals, as well as the many animal adaptations to floral forms. Butterflies and moths, for example, use coiled strawlike proboscides to sip nectar lying at the bottom of deep floral tubes.

It did not take curious naturalists long before they began to associate lepidopteran feeding behavior with certain shapes, colors, and sizes of flowers. Many languages have names for plants that essentially mean "butterfly-weed"—as is the case for a milkweed, *Asclepias tuberosa,* in temperate North America. Monarch and queen butterflies are among the many that frequent this orange-flowered perennial throughout its range. Vernacular Spanish lexicons from Latin America contain some wonderful associations between flower and pollinator, in which both are known by the same folk name. For example, *chuparosa* ("rose-sucker") refers not only to a hummingbird but to a red, tubular flower such as *Justicia californica* as well. It is true that hummingbirds in the Americas actively seek out tubular flowers that are often red or orange in color and rich in nectar. Hummingbirds learn that red signals the possible presence of nectar nearby. In fact, this color stimulus is so strong for hummingbirds that if you put on lipstick, fill your pursed mouth with wine, and stand where crimson-colored sugar feeders or potted magenta flowers have been outside, the hummingbirds will often feed right out of your mouth.

The premise that certain guilds of animals have strong allegiances to particular floral classes was first formally outlined by Frederico Delpino, a naturalist from Milan, in a series of papers published between 1868 and 1875. Paul Knuth, another giant in floral biology, is often credited for codifying pollination syndromes around 1910. These syndromes were later refined and amplified by the great German pollination biologist, Stefan Vogel, in a seminal essay published in 1954. In that essay, Vogel suggested that by recognizing certain patterns that he called *Stil* ("style"), floral biologists could predict which blossom visitors might be legitimate pollinators. Whether studied in the field or in the herbarium, Vogel argued that anyone could see how floral forms were suggestive of harmonious relationships with particular animals. And his essay illustrated these pollination syndromes with such beautiful plates that each relationship quickly became a truism in biological literature. To his credit, Vogel did caution his readers that such syndromes should be considered working hypotheses to be tested in the field. Nevertheless, other scientists took the color plates illustrating Vogel's classification at face value. For the two decades following his 1954 publication, biologists turned pollination syndromes into stereotypes, as if one-to-one relationships between plants and pollinators were the norm in the natural world.

Vogel's influence on correcting these simplifications was somewhat muted, for his publications appeared only in German. Other pioneering scientists such as Christian Sprengel, Paul Knuth, and Herman Mueller helped to establish the use of syndromes as a teaching tool in modern studies of floral biology, but their caveats were not heeded in more popular accounts. By the time Knut Faegri and Leendert van der Pijl's classic textbook *The Principles of Pollination Ecology* was published in 1979, the pollination syndrome was well established as the prevailing paradigm for studying the interactions between flowers and their pollinators. Although early workers intended that the concept be used experimentally, flexibly, it somehow became codified into dogma in the modern botanical literature. Too many traveling biologists became preoccupied with looking at the pollination syndrome classificatory approach as a fully explanatory signpost, and forgot to look for other signs and where they were pointing. Nevertheless, we believe that this methodology can still be used as an invaluable heuristic tool by students, naturalists, and biologists who aren't pollination biologists per se.

Certain floral shapes, fragrances, nutritional rewards, and blossom opening times became tightly linked to the evolutionary influences of a single species of beetle, bee, or bat. The lexicon for these "animal-loving flowers" currently in vogue among some biologists—*cantharophily, melittophily, chiropterophily*—is an outgrowth of earlier classification systems. Herbert Baker and colleagues elaborated these classifications to take into account not only floral morphology but nectar and pollen chemistry as well.

We know now that many of these interactions are not simple stories of intimate one-on-one relationships. In fact, University of California ecologist Nickolas Waser first encountered difficulties using such projections when he tried to explain the flowering times and seed set of the ocotillo's "hummingbird flowers" by the timing of hummingbird migrations through the arid Southwest. Waser and others were surprised to find that in many ocotillo stands, the red tubular flowers are nectar-robbed or pollinated by carpenter bees more often than by birds. Later in the summer, while desert bees are pollinating "hummingbird flowers," the same hummingbird that passed through the desert two weeks late for ocotillo may be pollinating 13 different kinds of alpine meadow plants, representing eight distinct plant families and several syndromes.

Studies by Liz Slauson of the Desert Botanical Garden in Phoenix, Arizona, on the floral biology and pollination of Palmer's agave *(Agave palmeri)* have also revealed a complex story of floral adaptations and guilds of primary and secondary pollinators. These massive century plants use a "big bang" reproductive strategy—that is, they bloom just once after storing up food reserves for decades, then the individual plant dies while its genes are carried into the next generation within the seeds shaken from the fruits. Their flowering stalks appear like ornate candelabras punctuating the fiery Arizona summer skies during sunset.

Although the flower stalk architecture is clearly modified (open and sturdy flower clusters presented on short horizontal branches), when night's darkness gives way to the chill of dawn, other interlopers are on the scene. At this time of the early morning, one can see and hear clouds of hungry bumblebees, honeybees, and carpenter bees along with paper wasps and giant orange and black tarantula hawk wasps. They greedily slurp up the bonanza of nectar and the female bees carry off the pollen. This is so because even in a site rich with local bats there is an over-abundance of floral goodies for all the players. But, as mentioned earlier, many sites are without the migratory nectar corridor-tracking bats or their numbers are much reduced due to bat roost destruction within caves or pesticide poisoning. Liz has determined that when bats are indeed scarce, the early morning bees get much of the agave pollen and nectar. Fortunately for Arizona century plants, bees are quite successful in pollinating the agave blossoms even though their floral evolution was most likely shaped by dancing with bats.

Elsewhere, in the lowland tropical forests, a single euglossine bee species may visit up to eight different kinds of orchids that flower sequentially in their Panamanian habitat. Waser and his colleague Mary Price have concluded that "generalization appears to be the rule among pollinators"—adding that for the various plants visited by hummingbirds at a particular site, "most of the flowers do not conform to a 'bird pollination syndrome.'"

Why then have some ecologists continued to use pollination syndromes as a tool for teaching about plant/animal interactions? The answer is simple. The syndrome game remains helpful as a point of departure when little is known about a plant's floral visitors. It may not suggest the entire range of animals to look for as potential pollinators, but it will help deduce a few of them. The pollination syndromes approach can still be used to advantage by students and other newcomers to the pollination game.

In some cases, we can only infer what the historic relationships between a plant and its predominant pollinators might have been, for these mutualisms have been profoundly disrupted by changes in the modern landscape. Such a deductive approach is routinely used by biologists visiting habitat fragments where only generalist pollinators such as honeybees persist. Floral morphology and nectar chemistry, for example, can be used to predict the historical presence of a pollinator—if that pollinator was singularly important in shaping the evolution of the plant's floral characteristics. And even though a plant may have gone extinct before its pollination ecology was adequately studied, the animals that once visited it may still show signs of adaptation to its kind of flowers. Moreover, certain inferences drawn from present-day fauna and from historic specimens may be used to confirm what the original interaction between plant and pollinator was like.

There are some remarkable cases of ecological detective work that resulted in the identification of pollinators other than those found visiting the plant today. In one case, plant ecologist Paul Cox was skeptical when he read secondhand accounts of rats pollinating an indigenous Hawaiian vine known as the ieie, *Freycinetia arborea*. It had few characteristics of a plant pollinated by nonflying mammals. He decided, therefore, to spend four days in a blind where some of the vines grew in forests above Kealakekua, Hawaii. During that time, he saw no rats on the plants and identified only one regular visitor to ieie flowers—a bird known as the Japanese white-eye. But the white-eye was first introduced to Hawaii from Japan in 1929, so its recent role in pollination of ieie flowers could not account for certain characteristics of the plant's inflorescences. The female flowering stalks of the ieie are rich in hexose sugars and certain amino acids that have independently evolved in the flowers of many plants attractive both to perching birds and to bats. Cox sensed that other birds must have been hidden in the history of the ieie vine.

Checking thousands of pages of Hawaiian naturalists' journals from the nineteenth century, Cox encountered notes suggesting that the islands were once visited by several birds that are now endangered or extinct: the Hawaiian crow, the oue, and a crossbill-like bird, last seen in 1894, that once lived on the Kona Coast. Cox took early museum specimens of these birds and began to examine their head feathers with a scanning electron microscope, searching for pollen grains. To the astonishment of his colleagues, Cox discovered dense quantities of a distinctive pollen type on the head feathers of the oue and Kona crossbill specimens and moderate loads of the same pollen on the head feathers of the Hawaiian crow. These pollen grains looked identical to those he obtained from a 57-year-old specimen of the ieie vine. Apparently these three birds regularly dipped their heads in the floral bracts of male ieie vines, and no doubt visited female flower stalks for other rewards as well. Cox's doubts were well founded: there was nothing about ieie flowers that suggested adaptations to rat pollination; in fact, that earlier claim was based on an error that had been magnified through repeated elaborations over the years. Instead, Cox proposed, the ieie flower stalk evolved for accessibility to a variety of birds, ranging in size from the introduced white-eye (a 4-inch-long bird) to the rare Hawaiian crow (almost 20 inches in length). The oue, the Hawaiian crow, and the Kona crossbill had all nearly gone extinct before the end of the nineteenth century, so the ieie must have suffered low pollination levels and seed set declines for a while. Nevertheless, these perennial vines could have persisted because they can vigorously reproduce by branching and rerooting. By the late 1930s, the introduced Japanese white-eyes had probably begun to take over pollination of the vine. Seed set no doubt rose again.

In another widely heralded case from Hawaii, a large number of "cardinal flowers" in the Lobeliaceae family have suffered extinctions in the last century, forcing their coadapted pollinators to shift to other floral resources. Of some 273 lobelioid species, subspecies, and varieties historically described from the Hawaiian Islands, only 27 percent have sizable enough populations to keep them from immediate extinction. A quarter of the species have gone extinct within the last century; another 19 percent of the species, subspecies, and varieties may be considered rare or endangered. This catastrophic decline in Hawaii's largest plant family poses the question: what has happened to their coevolved pollinators? For years, biologists had speculated that a number of Hawaiian birds known as honeycreepers may have coevolved with lobelioids, for some of them have long, downwardly curved bills that match the shape of floral tubes of certain cardinal flowers.

But when biologist Coleen Cory went searching for honeycreepers visiting the flowers of two rare lobelioids in the Koolau Range on Oahu, she failed to record a single bird on these plants during 136 hours of observation. A few hawkmoths, small black bees (the endemic bee genus *Hylaeus,* a colletid), and introduced honeybees were the only floral visitors she recorded on either lobelioid. She also found that these two rare plants are now capable of self-fertilization: they may not require birds or any other animal to move pollen from plant to plant in order to set seeds. Cory concluded that mutual codependence between honeycreepers and lobelioids was insupportable on the basis of present-day observations of both organisms on Oahu.

Recently, though, a team of zoologists has come to a different conclusion, arguing that the characteristics of living organisms cannot be understood in terms of modern conditions alone. When Thomas Smith, Leonard Freed, and colleagues looked at historic museum specimens of one Hawaiian honeycreeper, the i'iwi, they found that its bill was longer in the nineteenth century than it is today. Then they discovered a handful of historic notes from the nineteenth century. These notes documented i'iwi honeycreepers feeding on now rare or extinct lobelioids with long, tubular flowers that produced copious, hexose-rich nectar but no odor—floral traits that fit birds, rather than the insects Cory recorded. A long bill would have served the i'iwi honeycreeper well when these flowers were more abundant.

Most i'iwis have been observed feeding on the open flowers of the

*The nearly extinct i'iwi honeycreeper* (Vestaria coccinea), *from the Hawaiian archipelago, is the only known pollinator of several endemic flowers such as this Clermontia species. Its highly curved bill allows it to remove nectar efficiently from the similarly curved blossoms.*

ohia lehua *(Metrosideros)* tree, one that lacks tubular corollas specifically adapted to birds. Smith, Freed, and coworkers hypothesize that as the lobelioids declined in the nineteenth century, the once-common i'iwi honeycreepers began shifting to other floral resources in order to survive. Honeycreepers need no special bill adaptations to obtain nectar from ohia flowers. The zoologists suggest that the dietary shift from tubular flowers to open flowers resulted in *directional selection* for shorter bills. In other words: the i'iwis with the longest, most downward-curving bills were lost from breeding populations over the last hundred generations. The upper bills of i'iwis today are 2 to 3 percent shorter than those of i'iwis collected before 1902, when lobelioids were still quite common in the forests.

Such examples suggest that there are indeed long-term matches made between sets of pollinators and particular plants, but these matches are seldom so exclusive or inflexible that other organisms are permanently left out of the picture—whether those "others" are Japanese white-eyes or Hawaiian ohia trees. Unfortunately, many historic ecological studies tend to leave these other organisms out of the picture they frame by the way their goals are delineated. To overcome this tendency to look only at plant/animal pairs—rather than the entire interplay of floral resources and pollinators in a habitat—Judith Bronstein has encouraged us to take a step back and look at the entire landscape of ecological interactions revolving around pollen and nectar availability.

Admittedly Bronstein, like others before her, concedes that "we know remarkably little about how pollinators respond to the spatial and temporal variation in their floral resources." We do know that pollinators commonly move among patches of floral resources as they forage. And yet we seldom know how far they travel to make ends meet during a single day, let alone the distances they travel for food in their entire lifespan. Botanists at Barro Colorado Island, La Selva, and Monteverde in Costa Rica are among the few that have recorded exactly how many plant species are competing for the same pollinators in a given landscape—or, conversely, how plants have been selected to flower at different times and intensities to minimize competition with one another.

The most complete survey we know is a 50-month study of the insects that work the 133 kinds of plants found on 80 acres of Greek *phrygana,* a heavily grazed and burnt Mediterranean ecosystem of herbs and shrubs near Athens. In this disturbed community, which few ecologists would predict to be particularly rich, Petanidou and Ellis have documented the highest pollinator diversity yet recorded: an astonishing 666 species of insects, including over 225 species of solitary bees. Oddly, the climate is so variable at this site that only 20 percent of the entire pollinating fauna has been found present in all five years of the study. No bats or birds have been recorded in the area. But each flower in the *phrygana* averages five kinds of insect visitors, be they wasps, bees, butterflies, moths, beetles, or flies.

Gradually, Bronstein and others have begun to integrate various studies such as the Greek *phrygana* effort to give us an expanded view of plant/pollinator landscapes. Although additional studies now under way will add color and depth of field to these pictures, a few of these patterns can be illustrated here in a preliminary way. The first kind of landscape is dominated by generalist pollinators associated with flowers that bloom sequentially or without overlap. This kind of landscape is very common in the tropics, but it can also occur in highly seasonal environments all the way to the Arctic tundra. In tundra and in the wet tropics, the flowering times of a number of plants in the same habitat are complementary and do not necessarily compete for the same pollinators. In addition, the same plant may be visited by a sequence of pollinators over a long flowering season. For instance, lavender in southern Spain is visited by 70 different kinds of bees, butterflies, and moths, each with a different peak period of activity over a three-month blooming season.

At the Rocky Mountain Field Station in Colorado, Leslie Real and Nick Waser found evidence of *sequential mutualisms:* Flowering larkspurs supported hummingbirds soon after their arrival in the mountain meadows. The hummingbirds subsequently pollinated the late-flowering trumpets of *Ipomopsis.* Wherever these two flowers were found together, they were visited by the same sets of hummingbird species, but showed low levels of flowering overlap. On the floodplains of western Colorado below the Rocky Mountain Field Station, however, the Ute ladies' tress orchid demonstrates the perils of sequential mutualisms. Sedonia Sipes and Vince Tepedino have shown that because bumblebees rely on this rare orchid only for nectar, a sequence of other flowers must be available to provide it with pollen over the entire season. Lacking these foraging resources in the area, not enough bumblebees will stay to cross-pollinate the little orchid. Consequently, the presence of other species providing additional pollen sources for bumblebees may be just as critical as the pollinators themselves for the orchid.

Another example of this first landscape was recently described in forest fragments in Japan, where bumblebees sequentially visit a number of obligately outcrossing flowers over the span of a season. The earliest herb flowered in April, but did not have a good seed set, particularly where the forest fragments were found in residential areas. It appears that the season's bumblebee colonies had not yet grown large enough to service areas where there were insufficient flowers to attract them. When June came along, and the next herb flowered, there were ample rewards for bumblebees in the forest islands within agricultural and residential areas; the second herb had a high seed set in both areas. The third perennial herb flowered in August, and it too was not limited by pollinators.

The second kind of plant/pollinator landscape is dominated by generalist pollinators of plants that bloom all at the same time. This pattern often occurs in deserts and subtropical habitats where plants respond in concert to one brief rainy season, but it may be common in temperate and alpine zones as well. Plants in these landscapes are clearly competing for pollinators. And each may suffer low reproductive success, because bees or butterflies there may move indiscriminately from one species to the next, wasting pollen along the way. Some tropical herbs such as *Calathea occidentalis* produce unusually luxuriant floral shows to compete for the most effective pollinators, which in this case are bees that are seldom very abundant. More common butterfly visitors are far less reliable in moving pollen from one *Calathea* to the next, yet they clearly benefit from abundant nectar associated with these mixed species shows. Spring wildflowers along the Sicilian coast show a similar tendency toward profuse simultaneous blooms that attract a wide range of insect pollinators.

The third landscape pattern is dominated by specialist pollinators that visit plants having prolonged periods of flowering in an environment without seasons. Short-lived wasps, for example, must find a sequence of fig trees of the right species flowering in their local environment year-round, or else they will go extinct. Fortunately for the wasps, few fig trees in the same population flower at the same time.

In the fourth plant/pollinator landscape, generalist migratory pollinators such as nectarivorous bats switch through a variety of nectar-providing plants as they migrate from tropical to arid temperate environments. Even where there is only one nectar source available for a particular bat species, it usually flowers in sequence with other bat-

loving plants to the north and south, along the bat's migratory route. Thus the bat may specialize on just one kind of flower in each local environment, but these are linked into a *nectar corridor* of successive flowering times along the bat's migration route.

This concept of a nectar corridor for migrants, roughed out by Donna Howell in *National Geographic Research* in the mid-1970s, has recently been fully articulated and elaborated by Ted Fleming. His work has focused on the geographic sequence of plants used by lesser long-nosed bats—one of several pollinators of giant columnar cacti, century plants, and manfredas. These bats tend to utilize nectar from century plants in late fall and early spring and nectar from tree morning glories in the winter. In the summer, when they move great distances every month, they apparently seek out densely blooming patches of various century plants and columnar cactus species, foraging as much as 60 miles away from their day roosts. The migration of hummingbirds from northern Mexico through the Southwest over a monthlong period each spring apparently matches a similar sequence of flowering from south to north of various ocotillo species and populations. We have accompanied Ted on some of his nocturnal sojourns to the land of cardón and saguaro cacti alongside the Sea of Cortez near Bahia Kino in search of nectar-slurping migratory bats. Although bats are not the exclusive pollinators of either the giant columnar cacti or nearby century plants—remember, as Liz Slauson showed, that they can be adequately pollinated by bees and wasps in the morning after the blooms open—they do in fact appear to be pollen-limited at some sites and especially during years when bats are rare and visits to the flowers are infrequent.

The final landscape described by Bronstein is fairly rare: it is dominated by specialist pollinators, each associated with a small set of plants that flower at the same time. This landscape can be found in deserts, seasonal subtropical scrub, or alpine tundra. There we find dominance by oligolectic solitary ground-nesting bees, linked to a single widespread dominant plant such as creosote or mesquite, or moths that specialize in just one local yucca. Such insect specialists face serious problems if they emerge before or after the flowering of their sole set of floral resources—if late snows kill the earliest flowers, for example, or even postpone the entire flowering season. It can also occur in deserts following warm, frost-free winters that encourage the early opening of floral buds, weeks before migrating pollinators arrive. This is one more landscape where the perils of matchmaking are all too evident. Yuccas and yucca moths form one of the obligatory partnerships that occur between specialist pollinators and synchronous flowerers—and so they are vulnerable to a variety of destabilizing forces. It is no wonder that such one-on-one relationships comprise less than 1 percent of all observed plant/pollinator interactions in the environments that have been intensively surveyed.

But other plant/pollinator landscapes can become just as vulnerable as those with pollinators that specialize on synchronously flowering plants. If a specialist or a generalist dependent on sequential flowering of several species finds that one link in its chain is broken—through habitat destruction, say, or selective removal of the plant forming the most critical link—it may be unable to wait without food until other resources become available. In short, even pollinators that do not show strict dependence on a single flower may become vulnerable due to their dependence on a short list of floral resources.

More than 20 years ago, eminent tropical ecologist Dan Janzen became concerned with the vulnerabilities of such interactions as the dry subtropics of Costa Rica became more and more fragmented into a jumbled patchwork along the Pan American Highway. Janzen wrote prophetically about the perils to be faced by traplining hummingbirds that must forage along a 10-to-12-mile route through dry forest every day to obtain sufficient food for themselves and their young. When a significant portion of forest along their trapline is logged or converted to pasture, the entire trapline becomes worthless. Janzen wrote: "What escapes the eye . . . is a much more insidious kind of extinction: the extinction of ecological interactions."

While the few documented cases of truly obligate pollinators and synchronous flowers remain especially vulnerable, the other three plant/pollinator landscapes can also suffer from the extinction of mutualistic relationships. Yet each will occur according to its own pattern and at its own pace. It is easier to predict linked extinctions for strict mutualists which lack Japanese white-eyes or Ohia lehua trees to catch them in their fall toward extinction. However, other kinds of relationships can be profoundly disrupted as well. Just as we are beginning to turn our attention to plant/pollinator landscapes in their entirety, we are realizing that the big picture is rapidly becoming cracked, fissured, or torn to shreds.

*Bees come in all shapes and sizes, as shown in this selection of 11 species from the tiny* Perdita *to the giant carpenter bee* (Xylocopa) *in the center. Note the extremely long proboscis on the orchid bee* (Euglossa) *and the pollen loads on the bumblebee* (Bombus sonorus). *The bees are shown here approximately twice their actual size.*

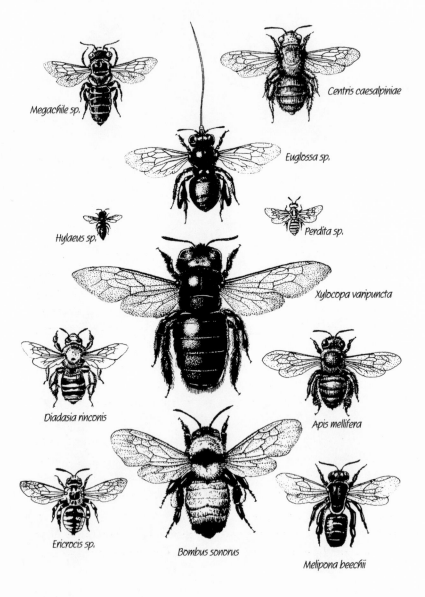

Megachile sp.

Centris caesalpiniae

Euglossa sp.

Hylaeus sp.

Perdita sp.

Xylocopa varipuncta

Diadasia rinconis

Apis mellifera

Ericrocis sp.

Bombus sonorus

Melipona beechii

# CHAPTER 5

# Bees in the Bestiary, Bats in the Belfry

## A Menagerie of Pollinators

*S*TEVE REMEMBERS:

I worked my way along the small desert wash set against Horned Lizard Hill, my eyes intent on bees darting into the canopies of flowering trees for nectar. The trees were ablaze with golden blooms here on the edge of Tucson, where Sonoran desertscrub vegetation furnishes food and shelter to a remarkable variety of native bees. I found them by the thousands—greedily drinking from all the brilliant yellow blossoms that had simultaneously exploded into flower. The woody legumes most attractive to the bees during mid-spring are formally named *Cercidium microphyllum,* but local folks know them as the little-leaf palo verde. Over a period seldom lasting longer than three weeks, each palo verde produces enormous masses of nectar and pollen attracting countless native bees per acre. A smattering of wasps, flies, true bugs, and beetles also find the lure of palo verde blossoms irresistible.

Within one palo verde canopy, I came across a 6-foot-tall saguaro cactus, putting forth an enormous white blossom for the first time in its 30 years of growth. A milkweed vine twined the two together—nurse tree and recently matured cactus—for a brief moment in time. Below the tree and the cactus were a dozen or so bellyflowers. Some had begun to bloom when they were no more than 4 inches tall and six weeks old. Most of them had recently passed their prime; their ephemeral flowers had withered and their seedpods were soon to pop open.

I took a swing at an extremely fast-flying gray blur of a bee, but missed. It was a male digger bee. It was an old friend to me: a big gray *Centris pallida* female, a harbinger of spring. In fact, I began studying the mating habits of this species with another friend, entomologist John Alcock, nearly 20 years ago.

In my head I began to refine the search image I needed to distinguish native solitary bees from the omnipresent honeybees on this and neighboring trees. I began to sort out the others by their flight patterns, their sizes, their colors. There were many digger bees, quite a few belonging to the leafcutter family, and even a few membrane or plasterer bees. Carpenter bees and bumblebees were active too. I then swept up as many kinds as I could capture that day, recording their nectar sources and periods of activity.

Back at the Arizona-Sonora Desert Museum at day's end, I dumped my catch onto a piece of notebook paper and sorted the pile of now quiescent bees into groups to show Gary. Each pile had bees that looked similar enough to one another to belong to the same genus or species. In all, during just a few morning hours, I had collected solitary and primitively social native bees belonging to 6 families, 20 genera, and perhaps as many as 50 different species. We identified representatives of *Anthophora, Ashmeadiella, Centris, Chalicodoma, Coelioxys, Diadasia, Ericrocis, Gaesischia, Megachile, Triepeolus,* and *Xylocopa*—just to rattle off some of the tongue-twisters. The number of species was more difficult to determine because a few of the harder-to-key variants may in fact be new species.

One thing was clear: Gary and I were living in the midst of a rich but imperfectly known Sonoran Desert pollinator fauna, including the bees. The possibility of nabbing something new to science—by intensively sweeping our nets through the canopies of even the most common trees—was not an unrealistic expectation for the days' bee hunt. Few bee biologists stay in one spot long enough to collect its full complement of pollinating animals. Even if they focus on just one

*On a southern Arizona hillside, a female carpenter bee* (Xylocopa californica arizo-
nensis) *is shown in flight returning to her nest gallery inside the dried stem of a sotol*
(Dasylirion wheeleri). *She has fashioned pollen and nectar from nearby mesquite and
ocotillo flowers into large provision masses for her voracious grublike brood.*

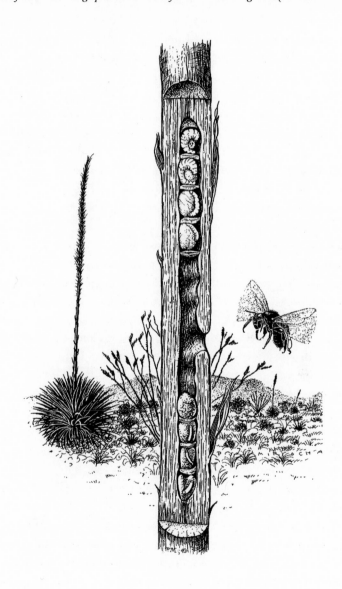

locality, few desert entomologists take the time to sample the entire blooming duration for even a few of the most dominant Sonoran Desert plants. What I could find might astonish my colleagues, for this desert is no barren wasteland as far as flowering plant and pollinator life are concerned.

Compared to other deserts around the world, the Sonoran Desert surrounding Tucson and extending into Mexico is exceedingly rich in native plants. On the 800 acres surrounding Horned Lizard Hill, our colleagues Jan Bowers and Ray Turner have identified nearly 350 kinds of plants. Gary and his students have collected another 366 species in two adjacent canyons 30 miles south, on the desert's edge near Tumaca-cori, Arizona. The Desert Museum's flora, including the adjacent Tucson Mountains, comprises more than 580 species of plants—and these plants may harbor almost 1,000 species of native bees in the same mountain range. That would make it the richest known parcel of bee real estate anywhere in the world.

Bees ignore several ecological axioms that are key determinants of diversity in other lifeforms. One such mandate is that as you advance north or south away from the tropical zones edging the equator, you will encounter fewer and fewer species of animals and plants. In other words: local species diversity for many kinds of organisms reaches a maximum in the lowland tropical regions, the zones lying within 30 degrees north and south of the equator. But with bees, this just isn't the case. With few exceptions, their species diversity and abundance decline as you move from the dry temperate to subtrop-ical regions in both hemispheres. Thus bees achieve their greatest abundance and diversity in the deserts and savannas of the world. It is estimated that some 1,500 bee species occur within the arid sub-tropical deserts and semi-arid uplands within an hour's drive from Tucson. In the wet tropical rainforests of the New World and Old World, however, bees are a distinctly underrepresented pollinator group. Tropical mountaintops seem to be peculiarly lacking in bee diversity. Such habitats nurture an astounding array of pollinators—flies, beetles, butterflies, moths, birds, bats, and other mammals that simply outnumber bees.

What they lack in number of species, however, the bees—especially the social bees living in populous colonies—make up for in overall biomass. There is one group of bees that is relatively abundant in the New World tropics: the fully social meliponines. Within that group, the stingless bee genera, *Melipona* and *Trigona,* are relatively abundant in wet tropical forests, especially in terms of the number of colonies per acre and their contribution to the overall pollinator biomass. These social insects are also exceedingly important as keystone members of such communities due to their activities: the "by-products" of pollination result in their cycling large amounts of nitrogen through these forests. Dave Roubik has estimated that 26 percent of all bees in the rainforests of French Guyana are stingless bees. Thus, while the overall species richness of bees seems to be reduced in the tropics, some groups of bees are disproportionately important.

Outside the tropics, bees are often more numerically abundant than most other taxonomic groups of pollinators, such as hummingbirds, bats, butterflies, or beetles. Within the continental United States, several entomologists have estimated that there are as few as 4,000 or as many as 5,000 species of native ground-nesting and twig-nesting bees. Other bee-rich continents or regions are those that have large expanses of deserts or savannas—such as Australia, where about 3,000 different kinds of bees occur, Africa, with over 3,000 species, and the Mediterranean countries, with about 3,000 species in their bee faunas.

Globally, taxonomists have named and cataloged about 25,000 bees and at least 4,000 wasp species active in pollination among the 103,000 known Hymenoptera species. This latter taxonomic group of bees and wasps contributes one out of every ten animal species now inhabiting the earth. Judging from the growth curve in newly discovered and named bees during recent decades, there are perhaps as many as 40,000 bee species alive today.

We cannot yet say why bees are distributed as they are today. Nor do we know which historical consequences of past plant distributions or competitive interactions between pollinators or nest site availability may have influenced where bees are found today. Even so, we can see a few general patterns from our viewpoint atop the desert mountain. For one thing, there are more bee species in regions with Mediterranean climates and vegetation: areas with warm, seasonally dry habitats that are often near the limits of subtropical regions. Bees are extremely abundant and diverse in the deserts of Israel, the southwestern United States, and northern Mexico, in Sonora.

The patchiness of floral resources—in space and in time—may be one driving force that has fostered the diversification of bee foraging strategies. Summer and winter–spring wildflowers require different sets of bees working the same spot of ground. Furthermore, different years' weather produces entirely different mixes of bees in dry locales. As noted in the Greek *phrygana* study mentioned earlier, only one-fifth of the entire insect fauna of this semi-arid site appeared, or was active, during each year of the study. Local bee diversity has also been influenced by the competition between bees and other flower-visiting creatures. When migrating hummingbirds miss the flowering of ocotillos, long-tongued or "robber bees" harvest a bumper crop of nectar.

As best we can tell, bee abundance depends on the availability of suitable materials—substrates—in which they can nest. Most bees are ground-nesters, so perhaps it is not too surprising that habitats having abundant bare ground—such as rocky deserts—have the greatest species diversity. Trampling by livestock can reduce the abundance of nest sites in semi-arid savannas and grasslands, but cattle are so sparsely distributed in true deserts that the rocky soils there still host a profusion of bees. Furthermore, since flooding on rocky desert slopes is a rare occurrence, underground bee cells are seldom inundated and ruined. Nor do soil-borne bacteria and fungi spoil their food stores as frequently as in the lowland rainforests. In sum, then, there are fewer constraints to bee diversification in the patchwork quilt of arid and semi-arid lands than in wetter habitats.

In a paper still widely discussed even though it was written a third of a century ago, the late, great ecologist G. Evelyn Hutchinson wondered out loud why so many kinds of animals could be alive in one place. Hutchinson intuitively turned to Santa Rosalia for help in answering this question, while visiting her mineral-encrusted bones in a grotto near Palermo, Sicily. There, at the shrine dedicated to the twelfth-century saint sheltered within Monte Pellegrino, he solved this old ecological riddle by reflecting upon a pool of water bugs. Hutchinson somehow recognized that the world's great zoological diversity was due to the interaction of certain forces. Wherever there was a mosaic of different physical and biotic microenvironments, diversity flourished. This mosaic was especially favored by small animals; in fact, the number of species increases directly with a decrease in the average body size of animals in a community. Deserts served as a mosaic of habitats, and clearly have triggered the *adaptive radiation* or species diversification of small pollinators such as bees. Each of these habitats had a different mix of floral attractants and rewards as well. As bees spread to different microenvironments to nest, they also developed ways to take better advantage of certain floral resources. Except for a few parasitic bees and a few "vulture bee" scavengers, all extant bees are true herbivores that eschew the taste of meat for vegetarian fare: they desire nectar and pollen. And as they began to specialize on a diet of liquid nectar and pollen grains, they developed new and often bizarre physiological and morphological adaptations.

Imagine floral nectars as their flight fuels and add to that a "bee bread" of protein- and lipid-rich pollen as brood food—this eccentric set of food preferences required the evolution of new toolkits by bees. Their tongues have taken different shapes suited to the job of sucking nectar from variously formed floral cups, bowls, and vases. As a result, bees sort out into two discrete groups: those with long tongues and those with short ones. The broad, often bilobed tongues of the short-tongued bees apparently came first. When deeply tubular corollas came on the scene, millions of years later, bees with longer tongues became more abundant. The bee's tongue is a marvelous organ, one that is not merely useful for the slurping of nectar. It is also a trowel for plastering saliva and other waxy or oily secretions to their burrow walls in order to keep their underground cell walls protected from moisture, fungi, and collapse. At the very tip of each bee tongue is a fringed spoonlike flap— the *flabellum*—that allows the tongue to rapidly lap up the nectar hidden in pools and crevices within nectaries of each flower.

Melittologists sometimes joke about bees as "flying Swiss Army knives," and for good reason. More than any other floral visitor—vertebrate or invertebrate—bees display bizarre structural adaptations on their legs for wildcrafting a living out of the floral world. Their front legs may hold crescent-shaped bottle openers. Their hind legs may have tibial spurs, pollen rakes, brushes, and baskets, as you can see on social bees like honeybees and bumblebees. Examine the hairs on the hind legs of different bees with a hand lens or dissecting microscope, and you will see strong differences in their arrangement, size, and texture. The bees that routinely buzz flowers to harvest pollen have very thin hairs that capture and hold the small dry grains. The squash and gourd bees of the American Southwest and Mexico, however, have coarser and widely spaced hairs that help them carry the massive oily pollen grains of cultivated squashes, gourds, and pumpkins.

Within the hundreds of orchid bee species, or euglossines, we see even more fanciful structural adaptations for handling floral offerings. Most orchids pollinated by these male bees play a game of deceit, offering bees neither nectar nor pollen as nutritious enticements for pollen transfer. The male bees do, however, harvest a floral fragrance from special patches of orchid tissue. They can do so thanks to their curiously modified forelegs, which have specialized hairs used for mopping over the scent patches of orchids. Thus the volatiles are wicked up by capillary action and later transferred from forelegs, to mid-legs, and finally to peculiarly swollen hind legs that hold reservoirs filled with spongy hairs. Here the males store up the orchid perfumes until they are needed as aphrodisiacs to attract female bees. The toolkits on their legs not only allow them to harvest sex scents from flowers, but also help the males work the perfumes into their own macho brew, an irresistible bee cologne.

Some of the most remarkable floral handling devices of all are found in bees of tropical climes. High in the rainforest canopies of Costa Rica and Panama, we have seen colorful, fast-flying anthophorid bees. Known as *Centris* and *Epicharis* bees, these beauties are in the oil business. On their forelegs are hairs that have been transformed during their evolutionary history into comblike squeegees. When these bees visit yellow flowers in the tropical family Malpighiaceae, they use the oil squeegees to rupture floral blisters containing novel and nutritious foodstuffs. The harvested oils are then transported back to the nest by holding them between the stout brown or black bristles on the bee's hind legs. These floral oils are then mixed with pollen collected from other kinds of plants to be offered as oily gruel to their developing young.

If that is not curious enough, try to imagine the "legs race" that has been occurring in the Cape region of South Africa for millennia. Here bees in the genus *Rediviva* have evolved very long front legs—in some cases, longer than their entire bodies. The legs may have increased in length over time in response to the floral evolution of their twin-spur snapdragon hosts in the genus *Diascia*. These snapdragon relatives have, over time, lengthened their floral spurs that house the energy-rich and nutritious lipids desired by these specialist bees. At the tips of *Diascia's* ludicrously long front legs, are, once again, those oil-sopping hairs.

Such toolkits have served bees well in their not-always-equitable "dances" with flowers. Call it a dance, or call it exploitation by both sides, pure and simple. Perhaps the term *floral reward* is a misnomer from earlier times when biologists believed these coevolved relationships to be one-to-one mutualisms that invariably provided equal benefits to both partners. Of course, not all flowers must provide massive concentrations of high-energy oils to feed their pollinators or hungry brood.


*S*TEVE REMEMBERS:
Some bees are so tiny that it's a wonder I even get to see them in flight. Fortunately, I have a few friends who are faster with a sweep net than I am. I was once in a field of low-growing matlike spurges, wondering what their minuscule flowers could ever muster to offer a pollinator. A moment later, I watched as one of my colleagues swept his net across the tops of the spurge mats, then thrust his head into the gossamer aerial net bag. Strange behavior? Yes, unless you are Jerome Rozen of the American Museum of Natural History, and you are seeking a rather diminutive bee. I walked up to Rozen, baffled, and peered at him through the fine-mesh weave of the net. "Looks like you didn't get anything much."

"You just can't see small bees," Rozen grumbled. He had, in fact, captured the tiniest of the tiny panurgine bees, known as *Perdita minima*. Its name alludes to the fact that the smallest of its kind stretch out to a mere 3/50 of an inch. It is the world's smallest bee, and it rivals a fig wasp and some thrips for being the world's smallest pollinator. And yet its pollination services are dependably provided to the matlike spurges which stuck that day to the soles of Rozen's feet. What the spurge provides in pollen and nectar to this bee would be starvation food for other kinds of pollinators. But for this lilliputian *Perdita,* it does the trick.

At the opposite extreme from the *Perdita minima* is a black giant, an immense leafcutter bee from the Moluccas islands halfway around the world. Alfred Russell Wallace, the codiscoverer of evolution by natural selection, spent time in these islands in the nineteenth century and found remarkable bees about 2 inches long living within the outer ramifications of termite "carton" nests. *Chalicodoma pluto* is the world's largest bee. This sexually dimorphic species has females with grotesquely enlarged mandibles and other mouthparts. They use custom tools to collect plant resins and transport them back to the nest. Until the late 1970s this amazing bee was known only from the type specimen in the British Museum of Natural History collected by Wallace, when it was rediscovered by Adam Messer, then a graduate student at the University of Georgia. You could balance dozens of *Perdita minima* upon the antennae of Wallace's lost bee.

Of course, there are many pollinating insects smaller than the smallest bee. There are over 500 species of flower thrips active in pollination, but they are not typically oriented toward pollen transport from one flower to another of the same species. They are, however, effective pollinators of many flowers. In fact, the diminutive thrips have recently been demonstrated to be highly effective obligate pollinators of certain dipterocarp canopy trees in the rainforests of Indonesia. Thrips typically range in size between 0.04 and 0.12 inch in length.

In this chapter, we can only glimpse some of the bewildering and fascinating shapes, sizes, and behavior of the myriad insects that visit and pollinate flowers. To fully do justice to insect pollinators other than bees, we would need pages to explore each major group of pollinating insects (beetles, butterflies and moths, flies and wasps) merely to highlight their biodiversity, their relative importance to pollination, and their astonishing lifestyles. Nevertheless, we offer a few case histories here to place these organisms in a conceptual framework within the vast worldwide pollinator bestiary. Ever since their origins more than 100 million years before the present, various families and genera of the largest and most successful insect orders (*Coleoptera, Diptera, Hymenoptera,* and *Lepidoptera*) have diversified to exploit a wide range of flowering plants utilizing varied strategies. Some specialize on a few related plants; others are generalists that draw on a wider dietary base.

If we were to kneel among the brilliantly colored and fragrant wild-flowers of an alpine meadow, our attention would soon be diverted by the guests invited to feed at the banquet. The air is filled with thousands of flying insects of all possible sizes, colors, and forms. The combined noise of their beating wings is especially loud; the sound from a low-pitched passing bumblebee careens past our heads. Smaller insects are everywhere—from tiny straw-colored thrips invisibly feasting upon pollen inside flowers, to acrobatic flower flies, to bee flies, buprestid beetles, spider wasps, bees, and a winged gallery of gaudy butterflies.

All of these insects are floral visitors, but not all will acquire pollen that can be passed along to the next flower on their visits. Some have come to slit "floral throats," robbing them of sweet nectar or stealing away with pollen without fulfilling the implied pact with the flowers. A few of the insects—especially bees, flies, and butterflies—are excellent and faithful pollinators upon which the local flora "entrusts" its same-day pollen delivery service. Let's depart now from the flowers of the meadow and examine the diversity of these pollinators.

More than all the rest combined, the order Coleoptera (with over 350,000 named species worldwide and many yet to be discovered) is the largest extant insect order and probably always was so. From the sap beetle pollinators of western spicebush to the specialized scarab beetles that enter and pollinate the flowers of the giant Amazonian waterlilies, beetles are the customers and pollen vectors of choice for thousands of flowering plants on most continents.

Approximately 30 families of beetles are today engaged in the polli-nation trade, often acting as what has been termed "mess-and-soil polli-nators." While the label is not terribly flattering to this ancient lineage of inordinately successful insects, it does indicate their mode of entry and gustatory pursuits. Thus, sap and rove beetles attracted to the fra-grance of the western spicebush merrily chomp on special food tissues and on modified petals, in addition to the pollen grains. In so doing—and amidst a good deal of copulating and defecating—they effectively move the spicebush's and their own gametes around.

Whenever a bloodthirsty female mosquito peskily buzzes around our heads in a darkened room, we aren't likely to thank the males of its species for the pollination of rare orchids in Wisconsin peat bogs. Yet male mosquitoes seek out nectar-producing orchids and other plants and are dependable pollinators in many parts of the world. In temperate alpine meadows, there are often dozens of species clambering over the open, broad clusters of blossoms on plants like the giant cow parsnip. Our attention is diverted by the high-pitched whine and darting motions of a fuzzy golden bee fly with a black beaklike set of mouthparts used for extracting nectar from nearby blossoms. Thus the flies are exceedingly diverse and important pollinators the worlds over. The order to which they belong, the Diptera, contains over 150,000 described species. And of those species with a taste for food on the half petal, there are at least 45 families of flies that routinely visit flowers.

Tubular flowers that are often pink or yellow in color with a sweet scent and abundant nectar at their base attract those scaly winged beauties sought out by "butterfliers" (a new breed of butterfly hunter who do their hunting with binoculars, notebook, and pencil). Butterflies are active by day and are found in about 16 families that regularly visit flowers in search of nectar. The order to which moths and butterflies belong, the Lepidoptera, contains at least 100,000 living species according to current estimates by modern taxonomists. It may surprise the nonentomologist to learn that moths, the butterfly's nocturnal cousins (actually butterflies are likely derived evolutionarily from distant moth ancestors), outnumber the butterflies by about ten to one. And yet moths are extremely important pollinators of night bloomers including the sacred datura and many cacti.

Although not so numerous as bees, their "colleagues" in the order Hymenoptera, wasps, also pollinate certain flowers. In the American Southwest, many spider wasps (like the giant tarantula hawk) are important floral visitors and pollinators of native milkweed plants. Similarly, figwort blossoms are especially adapted for visitations by wasps. Many wasps have bodies that are too smooth—especially when compared to their hairy cousins the bees—to pick up much pollen. Some wasps do, however, have legs with coarse hairs that are adequate for picking up and transferring pollen from flower to flower while they go about their business of searching for sweet nectar within blossoms. There are about 10 to 15,000 species of wasps that do function to some degree as pollinators of flowering plants.

It's hard to decide who might be the largest of the world's vertebrate pollinators—not because we don't know which floral visitor is the biggest of them all, but because we don't know which dependably moves pollen between different flowers. Our guess, however, is that this distinction belongs to a Malagasy lemur, now highly threatened. Compared to the tiniest bee *(Perdita minima)* or a thrip, the black and white ruffed lemur is a thousand times larger from its head to the tip of its tail—in all, some 4 feet in length. It is the largest of the quadrupedal lemurs still left on this planet. This lemur is also many times more endangered than the tiniest bee, not only because of continued destruction of Madagascar's lowland rainforests, but also because it is hunted and trapped as a human food delicacy throughout its restricted range. Between 1,000 and 10,000 of these lemurs remain in the wild. Nearly 500 are held captive in a hundred different zoos around the world. The subspecies *Varecia variegata variegata* is considered to be endangered according to World Conservation Union criteria. Conservation International has given its conservation and habitat protection a "high priority" ranking, since it remains a popular target for Malagasy hunters.

In a recently celebrated confirmation that large, nonflying mammals can be effective pollinators, Hilary Morland of the Wildlife Conservation Society spent several seasons watching 10-pound black and white ruffed lemurs. These lemurs lived in a forest of "traveler's trees," *Ravenala madagascarensis.* The national tree of Madagascar, this extraordinary plant has a single palmlike trunk which may grow 100 feet tall before sprouting a single vertical fan of banana-like leaves. During 40 daylight hours of observation, the lemurs were seen ascending the trunk to make 57 visits to the pale yellow flowers of the traveler's tree. Once the lemurs arrive at a flower stalk, they use their nimble hands to pull open the tough bracts protecting a dozen or so flowers, then stuff their muzzles inside each flower to drink its nectar. After fruit, the traveler's tree nectar is this lemur's most important food. No other vertebrate in Madagascar appears to have the combination of agility and strength required to open the bracts to obtain the floral nectar of the traveler's tree. Morland and colleagues confirmed beyond doubt that this 10-pound lemur carries pollen in its fur from one plant to the next, and that its association with traveler's trees is ancient.

Of course, length from head to tail is not the only way to determine maximum size of a pollinator. Wingspan is another, and some of the flying foxes open their forearms to let their wings cover a 5 1/2-foot breadth. The largest flying foxes, however, are fruit-eaters first and come in a poor second as flower visitors and nectar drinkers. Their incidental movement of pollen, moreover, is not always well targeted. But certain smaller flying foxes in the genus *Pteropus* are not the wasters of flowers that their overgrown kin may be. *Pteropus* includes 56 to 59 species occurring east from the islands in the Indian Ocean, well into the islands of the South Pacific. In fact, some have been found 200 miles out at sea, away from any landmass, so it is conceivable that flying foxes may actually be able to move pollen some distance between islands.

On the islands and peninsulas that flying foxes frequent, there is often a paucity of other vertebrate pollinators. Indeed, many plants rely solely on flying foxes for transporting pollen from plant to plant. Paul Cox and colleagues report that more than 92 genera of plants in 50 different families have been visited by flying foxes. Unfortunately for the plants that rely on this keystone species, many populations of flying foxes have suffered declines nearly as dramatic as those of lemurs. In the Philippines, where 150,000 flying foxes would congregate in the 1920s, the largest gatherings today are seldom more than a couple hundred individuals. Three Pacific Island flying foxes have already gone extinct. Twelve other species are of concern to IUCN, and the U.S. Fish and Wildlife Service lists three species as endangered on Pacific Islands claimed by the U.S. government.

In many ways, the North American equivalents of flying foxes are the nectar-feeders among the American leaf-nosed bat family, the Phyllostomidae. Thirteen genera in the subfamily Glossophaginae have been confirmed as pollinators for a variety of plants, ranging from bananas and tree morning glories to towering century plants and columnar cacti. Mexico alone has 11 species of nectar-feeders in seven different genera, and 6 of those species are found nowhere else except Mexico. Two of them are locally rare, and there are unconfirmed reports that several others are in decline.

Aside from flying foxes and nectar-feeding leaf-nosed bats, 15 other genera of bats pollinate plants on various continents. Collectively they include at least 75 additional species that feed on nectar or pollen of vascular plants, including some plants known to be rare. It appears that a disproportionately large percentage of the 56 endangered species in the Chiropteran (bat) order are nectar-feeders and pollinators—also true of the ten bats that have already gone extinct. Nectar-feeding bats are a large component of the 533 mammal species considered threatened with extinction by the Global Biodiversity Assessment.

Other, nonflying mammals are reported to have pollinated certain plants, but most of these reports are anecdotal at best. Our colleagues Charlie Jansen and John Terborgh have done much to establish that nonflying mammals such as opossums, marmosets, and tamarins are legitimate pollinators in the neotropical forests, and their rigorous methodology will no doubt be used by others to add species to pollinator lists. But the accepted cases of nonflying mammals serving as pollinators come mostly from Australia, where honey possums, dibblers, dormouse possums, feather-tailed possums, pygmy gliders, brush-tailed possums, and spotted cuscuses are among the marsupials that regularly feed on flowers. There are also reports of pollination by tree squirrels, bush rats, galagos, tree shrews, raccoons, kinkajous, olingos, and long-tailed weasels. Many of these mammalian visitors destroy flowers while obtaining nectar, however, and spend an inordinate amount of time in single trees, rather than transferring pollen from one plant to the next. If raccoons are indeed found to be legitimate pollinators of flowering plants, they will become the Heavyweight Pollinator Champions of the World, for they may weigh more than two and a half times the weight of the fattest variegated lemur.

Mammals may be the biggest vertebrates that serve as effective pollinators of plants, but they are hardly as diverse as the avian pollinators worldwide. Over 1,500 bird species in at least 18 families have been confirmed as effective pollinators of plants. Common names such as honeyeaters, honeycreepers, flowerpeckers, and honeyguides hint at these species' reliance on nectar. They range in size from 2-inch hummingbirds to Hawaiian crows ten times their size.

Of all the continents, perhaps Australia has the most widespread occurrence of birds as pollinators. More than 110 species of birds have been seen visiting some 250 plant species in Australia alone. Some 70 species of honeyeaters may be effective pollinators of Australian plants. Lorikeets, parrots, silver-eyes, woodswallows, chats, sunbirds, orioles, trillers, thornbills, shrike-thrushes, treecreepers, bowerbirds, and butcherbirds have also been seen taking pollen or nectar from Australian flowers.

It is not surprising that honeyeaters and certain other nectar-feeding birds have brush-tongues that mirror the morphological adaptations to nectar sucking found in certain bees. Theirs too presumably evolved for licking up sticky nectar. Asian flowerpeckers and honeyguides, Hawaiian honeycreepers, African sugarbirds, Australasian honeyeaters, paleotropical white-eyes—all show similar adaptations to particular floral traits such as long tubes and copious daytime nectar production. Neotropical hummingbirds clearly have no monopoly on avian adaptations to floral rewards.

At least 42 genera of nectar-feeding birds of the world, however, now include species threatened by the loss of floral resources and nesting habitat. Among these are no fewer than 26 hummingbirds considered globally threatened. Some, such as the hook-billed hermit and Chilean woodstar, are clearly endangered by the diminution of nectar resources resulting from massive tropical forest conversion to croplands. In addition, ornithologists remain concerned about 22 other hermits, coquettes, pufflegs, and metaltails in the hummingbird family, as well as 37 white-eyes, 7 flowerpeckers, 11 honeyeaters, 4 honeyguides, and 7 orioles.

Although fish can walk and a few glide and fly, none have been reported as pollinators in the journals that we regularly read. The same can be said for frogs and salamanders. Thus the global decline in amphibians will not directly affect seed set among flowering plants. But a single reptile has made it onto our list of pollinators, as if to remind us that the Natural World seldom says "never." Geckos are the sole group of lizards that have been caught in the act of pollination—prying open the long flowers of New Zealand flax and inserting their tongues in the floral tubes to suck up nectar. These remarkable reptiles live on a few of New Zealand's offshore islands but are highly endangered due to introduced birds and mammals. While nectar-feeding, these geckos often brush up against flax anthers in a way that leaves plenty of golden pollen on their chins and throats. In fact, their throat scales are modified to hold onto flax pollen grains—much as the hairs of the lesser long-nosed bat in Arizona are elaborated to increase their surface area for effective pollen pickup and delivery. The geckos then move on, carrying some of this pollen with them to another flower, sometimes on a plant located a considerable distance away from the first.

All told, we conservatively estimate there may be between 130,000 and 200,000 invertebrate and vertebrate species that regularly visit the flowers of those higher plants which depend on animals to assure cross-pollination. This number of animals is at least half the magnitude as the number of flowering plants (other than grasses) described in the floras of the continents of the world. How many are dependable, effective pollinators remains to be seen. And only our grandchildren will be able to grasp how many of these animals and plants survive the next 50 years, for the biological diversity of the entire planet is facing unprecedented threats.

*At dusk, a giant sphinx moth* (Manduca quinquemaculata) *hovers just above the trumpetlike and sweetly
scented blossoms of jimsonweed* (Datura discolor).
*Note the long proboscis — the tongue — slightly bent
and inserted deep within the flora tube to sip nectar.*

# CHAPTER 6

# Fractured Fairy Tales

## Disruptions in Fragmented Habitats

*G*ARY REMEMBERS:

It was a dream of a midsummer night for desert ecologists, one that should have appealed to moths as well. Two kinds of nocturnal bloomers were breaking bud, ready to blossom and release their intoxicating fragrances. I was out in the Sonoran Desert along the U.S./Mexico border—a protected area on one side of me, a scatter of cotton fields, cattle ranches, and garbage dumps on the other. Along with Mexican ecologist Humberto Suzan, I had been tagging and monitoring a rare night-blooming cereus cactus for well over a year, and we hoped this night would reward us for our efforts.

Although I knew where every flowering cactus could be found over several hundred acres, I had yet to see many flowers, for *Peniocereus striatus* populations bloom in just a few episodic spurts each year. With volunteers from Mexican and U.S. universities situated next to flowers on both sides of the international boundary, we were hoping to catch sight of the moths presumed to be the pollen vectors that could keep this binational population a cohesive whole.

The only variable we had not pinned down was the local availability of two rather widespread hawkmoths: the white-lined sphinx, *Hyles lineata*, and the tomato hornworm, *Manduca quinquemaculata*. I had begun to worry about the moths earlier in the day when visiting the study area in preparation for the night shift. While tagging soon-to-open floral buds, I noticed a cropduster flying straight toward me, arriving to spray insecticides in the cotton fields on the Mexican side of the cactus patch. As the plane made its first pass over the fields to kill any larvae feeding on the cotton bolls and foliage, it kept its nozzles open far too long: insecticides were sprayed past the field's edge, into the protected area on the other side of the boundary line. The chemical mist could be seen settling into the places where the cactus grew in between the cotton fields, and into the denser trees within the protected area as well.

The same day at dusk, Humberto and I went to sweep for hawkmoth pollinators in a weedy row of datura surviving on the edge of the cotton field. Its flowers had opened in advance of the cactus flowers, so it was already attracting the last of the afternoon's bees and the first of the evening's hawkmoths. Some claim that datura or "jimsonweed" flowers offer moths an alkaloid-laced nectar capable of intoxicating them, a nectar even more attractive than that of cactus blossoms. We set out to see if this was true. Every fifteen minutes, from 6:45 in the afternoon to well after 8:00 P.M., Humberto and I walked the row of jimsonweed, sweeping our nets over all opened flowers, counting the number of moths per period.

The results of our sweeps—on this night and nine others—were disheartening. Amidst irrigated weeds that might otherwise have supported greater densities of moths than the desert itself, the hawkmoths were few and far between this year. Why? The local farmer hinted at the reason. He had called the cropduster over to his fields to spray every eight days, thanks to a Mexican pest control program subsidized by a U.S. federal agency. This frequency of spraying had knocked back the densities of most moth larvae feeding on cotton and other crops nearby, not to mention those adults which favor jimsonweed and cactus blossoms. Thirty-two hundred gallons of pesticide are typically sprayed over this valley's 2,000 acres of crops, most of them aimed at the pests in the 350 acres planted to cotton.

As darkness fell and temperatures plummeted as well, the cereus cactus flowers opened. Each of our volunteers had snuggled up to a spindly plant studded with a bunch of trumpetlike flowers, but few of them heard a moth flying during the next few hours. Dozens of scarab beetles came to rob nectar from the flowers, but only two hornworm moths were spotted the entire night.

Our season's cumulative counts of flower buds, opened flowers, and ripened fruits suggested that the borderline cereus were indeed pollinator-limited during the summer of 1991. Of some 64 floral buds that were initially produced, only 27 percent were pollinated and matured into ripe, red fruit. In some of the most isolated but spray-vulnerable fragments of the borderline population, fruit set was as low as 5 percent of the buds produced. And even though a fully pollinated cereus cactus fruit can contain as many as 360 seeds, most of the mature cactus fruits we dissected held less than 100 seeds. One had only 48 seeds, indicating that less than a seventh of the ovules were fertilized.

These counts were telling signs of pollinator limitation. We guessed that the cactus required cross-pollination to set seed, but its pollen was not being moved frequently enough between plants. Of the few hand pollinations we made to see if the cereus could be self-pollinated, all of them aborted, suggesting that the flowers were not self-compatible. At another site far away from agricultural pesticide use—where pollinators were abundant—one plant produced 24 open-pollinated flowers and 21 of them ripened into fruit. This plant obviously had ample stored food resources for high fruit set to occur if enough pollen had been delivered to its waiting flowers by pollinators.

All signs pointed to a dovetailing of two problems. First, human conversion of desert habitat to fields, stockyards, and houses had reduced the cactus patch size in this borderline landscape. Then insecticides and other chemicals had reduced the numbers of pollinators available to connect the scattered survivors. After a while, we began to refer to the cereus dilemma as "chemically induced habitat fragmentation," as if a multisyllabic diagnosis might cure the desert's woes. In fact, the cactus habitat was disrupted more than any map could reveal. Even in the protected area where natural desert vegetation appeared to be intact, insecticide spray drift had drastically reduced the frequency of moth visits to cactus flowers.

Alongside the fields, herbicides were periodically applied to kill back the jimsonweed, a plant otherwise so abundant that it tended to keep moths available for cereus blossoms, since it bloomed many more nights each summer than the cactus did. Jimsonweed is also a larval host plant for the hawkmoths, as are devil's claw and coyote tobacco, two other weeds that grow in and near irrigated fields. The local farmers, however, have none of the great love for these plants that their O'odham Indian predecessors in the region have. Year in and year out, weeds are sprayed, scraped, and hacked away as soon as farmworkers can get to them. While herbicides and hoes seldom reached into the cereus cactus patches, we still saw little evidence of recent recruitment among these night-bloomers. Rabbits and other browsers run rampant around the irrigated fields, and often nip away at any new cactus recruits they find on the desert hills above the fields. In addition, woodcutters have been coming to the hills to chop away at the mesquite and ironwood trees where the cereus hide themselves. Many of them, therefore, are left without protective cover, fully exposed to the baking sun of summer, the blistering freezes of winter, and the trampling hooves of cattle year-round.

Still, we were not fully convinced of the dovetailing between land conversion and chemical inundation until we studied the Sonoran cereus in an area relatively free from both effects. More than 200 miles south along the Sea of Cortez coast, we encountered another population of the same species of cactus in an area far removed from any agricultural damage. Although it was much drier than the cactus patch along the border—and with less jimsonweed available—the density of moths we observed was at least eight times higher along the desert coast.

There we brushed fluorescent dyes into opening flowers, then watched as visiting moths punched their proboscides down past the dye-and-pollen-laden anthers in search of nectar. We then carried around black lights to find where else the moths had visited—and where they had deposited the fluorescent dust along with pollen. As we ranged farther and farther from the flower we had initially dusted with dye powder, we were amazed that the frequency of other flowers receiving dust did not immediately slack off. More than 250 yards from "home base," we were still finding ample traces that hawkmoths had been busy cross-pollinating the cereus flowers. These rustic indicators buoyed our confidence that along the Sea of Cortez coast, cereus plants would achieve high levels of seed set and their gene pools would not stagnate.

When we returned from the seashore to the border, it was with a certain sadness. The United States had established a protected area along the border more than 50 years ago without knowing that a rare cactus occurred within this land reserve and few other places north of Mexico. Once they discovered it, though, rangers considered its status essentially safe so long as cactus collectors could not track it down or dig it up (which is why I have avoided specific place names here). What they did not foresee was that establishment of the protected area would stimulate the growth of the farming pueblo on the other side of the line. Restaurants were erected to serve American tourists, and they needed meat, vegetables, oil, and fuelwood to keep their customers well-fed. So fields were cleared right up to the edge of the land reserve, and firewood was cut on the sly even within its boundaries. Pesticides, herbicides, exotic grasses, and rabbits crept into the land reserve, especially in the vicinity of the night-bloomers.

Belatedly we realized that the cactus plants could not be considered safe simply because they were nominally protected within the U.S. reserve area. There were far too few of them north of the border to produce sufficient pollen to saturate every nearby stigma. In effect, they required pollen flow from across the border to achieve adequate cross-pollination for seed set. Yet their fast-flying pollinators were being locally depleted. The population fragment that survived in the land reserve would continue to decline so long as recruitment of new seedlings failed to equal the number of plants succumbing to a variety of natural and human-caused stresses. The mere presence of the cactus plant within the boundaries of a protected area was not enough to ensure its survival.

If the Sonoran Desert borderlands were the only place in the world where habitat fragmentation has affected rare plants, we would hardly have cause for concern. But similar reports have been arriving in recent years from around the world. Izumi Washitani's careful work with a Japanese primrose, *Primula sieboldii,* presents the most dramatic case of pollinator loss due to habitat fragmentation caused by urban sprawl. As Washitani and his colleagues have recently described, this perennial herb originally occurred

> in a range of moist habitats throughout Japan, but in recent years [it] has become endangered due to habitat destruction by human activities. In particular, its lowland floodplain habitats have been almost completely destroyed.... [However, one 10-acre area] of the moist tall grassland community in the floodplain of the Arakawa River in Urawa City adjacent to northwest Tokyo has been enclosed as a natural reserve since 1961. This site is inhabited by a remnant population of *P. sieboldii.* [It] suffers from severe limitation of pollinator availability because the reserve is surrounded by urban areas and golf courses which limit suitable habitats for pollinators.

When we recently interviewed Japanese environmentalist-poet Nanao Sakaki about the primrose population, he likened it to a bonsai: "The plant [population] has been shrinking . . . what's left is so small in scale in Japan. Around it, water was muddy, polluted, full of pesticides, dirty dish water. The primrose was lovely, once, a carpet." Now it is just a throw rug.

At the height of the flowering season, Washitani observed not a single insect visitor to 68 flowers over 16 hours of constant attention. The entire insect fauna of the reserve appears impoverished, but there is a particularly conspicuous lack of bumblebees, the most active pollinator of other primrose populations on the nearby island of Hokkaido. There, bumblebee queens still transfer pollen between the different-shaped pin and thrum flowers, ensuring high seed set. In the remnant habitat along the Arakawa River, however, seed set is not only more variable: it is negligible or altogether absent on some of the plants.

The bumblebees may require pollen and nectar from additional forage when the primrose is not flowering, but these resources may not be available in the reserve. Moreover, insecticide use on the adjacent golf courses, or trampling of nest sites in the reserve, may have harmed the bumblebees. The primroses of Tajimagahara Reserve will simply not survive very long if their pollinators need other resources that cannot be found in the remnant habitat retaining the rare plant. This is why Daniel Janzen reminds us of "the eternal external threats" that can undermine the regenerative capacity of plants and animals in small reserves. Until we deal with the pressures emanating from beyond the boundaries of protected areas, we will continue to witness low reproductive success and plant population declines in these areas.

Of course, not all pollinators are affected by habitat fragmentation in the same ways, nor at the same rates. In northern Europe, Jennersten has studied wild habitat fragmentation in landscapes dominated by modernized farms to observe its effects on bee visitation to flowers of rare plants and the resulting rates of seed set. He has found that short-tongued bumblebees seem to be more rapidly affected than long-tongued bumblebees by the reduced patch size of certain rare flowers. Although Jennersten's hand pollinations quadrupled the seed set of rare wildflowers from the smallest patches of habitat, they made much less difference in larger populations that had hardly been damaged by industrialized agriculture. In the islandlike habitat patches where only a few wildflowers persisted, their period of nectar and pollen production was too brief to maintain a diversity of resident pollinators, which require fuel over a long active season. The short-tongued bumblebees were among the first to go looking elsewhere for resources when patch sizes got too small.

The phenomenon of disappearing bees and diminishing plant reproduction has attracted enough attention that it has been given a name: the Allee effect. When a population's size drops below a certain threshold, it can no longer support its ecological associates and it will lose its viability. First elucidated among the microscopic organisms known as rotifers by the pioneering ecologist W. C. Allee, the principle clearly applies to rare plants that cannot sustain pollinators or seed dispersers necessary for the regeneration of their populations. Unless there is rapid selection for rare mutants that allow some other form of reproduction—self-pollination as occurs in peanuts or vegetative cloning like potatoes—the population of plants will die out once its attendant pollinating animals have been removed.

If you wish to watch a plant species getting knocked out, blow-by-blow, by the sheer pressure of habitat fragmentation, there may be no better place for a ringside seat than in Western Australia. In an area where farming and road building have rapidly fractured semi-arid habitats over the last century, there remain only 16 small populations of a cone-bearing shrub known as Good's banksia. Nine of the remaining populations average only eight plants per patch. The larger populations average 150 plants, but they are isolated within farmsteads or in nature reserves. Yet even among the largest populations, seed production is no longer reaching its maximum potential. This is not because poor soils or increased competition are reducing the number of cones that a plant could produce. Instead, it looks as though the cones are not being fully fertilized.

The smaller populations clearly have lower levels of seed set, even though the shrubs are roughly the same size as those in the protected populations. In five of the nine smallest populations each covering less than 200 square yards ecologists from nearby Perth discovered that no fertile cones had been set for a decade. These roadside populations are all too small to attract pollination activity by nocturnal honey possums or by honeyeaters, which are most active at daybreak. Although the shrubs are situated along roads that receive hardly any regular vehicular traffic during the night or at daybreak, the pollinators are nowhere to be seen.

B. B. LaMont, one of Western Australia's experts on disappearing banksias, has concluded that the value of the roadside populations in conserving Good's banksia as a species is negligible—the smallest patches are functionally extinct, for they lack any means of regeneration. He and his colleagues do not mince their words: "The sobering message for land managers and population modellers is that a series of small populations may not have the same conservation value as a larger one with the same total population size. . . . [There] may be a threshold

*In Australia, a small marsupial known as the sugar glider* (Petaurus breviceps) *crawls among the conelike blossoms of a banksia* (Banksia burdettii), *a member of the family Proteaceae. These mammals, along with certain nectar-feeding birds, are important pollinators of this large plant genus.*

below which local extinction is inevitable." Beverly Rathcke and Eric Jules of the University of Michigan echo these concerns:

> Habitat fragmentation immediately reduces the sizes of species populations, increases their isolation, surrounds them in a matrix consisting of a new environment such as agricultural fields or [urban] development, and commonly changes their environment. . . . [It] is often considered to be one of the greatest threats to terrestrial biodiversity. . . . All available evidence shows that pollinator abundance and diversity decline with fragmentation.

While habitat fragmentation certainly affects the reproduction of rare plant populations, it is not always clear what their original pollinators were or what forces caused the pollinators' demise. California botanist Bruce Pavlik has brooded over this dilemma with regard to the Antioch Dunes evening primrose, a rare subspecies now restricted to less than 12 acres of sand along the confluence of the Sacramento and San Joaquin rivers. The few acres of sand supporting this evening primrose—along with the Contra Costa wallflower and a Lange's metalmark butterfly—were awarded the first "critical habitat" designation after their federal listing under the Endangered Species Act in 1973. And yet, designation of this protected area may have come too late to prevent the loss of pollinators for the evening primrose known as *Oenothera deltoides* subspecies *howellii*.

G ARY REMEMBERS:
When I visited the Antioch Dunes in the spring of 1995—after years of reading Pavlik's work—I was emotionally unprepared for the pitiful condition of the plant populations remaining there. I could not believe the extent to which they were surrounded by factories, power plants, and railroad yards. It seemed to me that they were overwhelmed by exotic weeds such as brome grass, Eurasian oats, yellow star thistle, and wild cucumber vines. I heard the constant roar of heavy machinery coming from the adjacent gypsum plant. Beneath high-voltage power lines running from one huge steel tower to the next, a wooly plant with gray, crinkled-up leaves and huge white flowers sat low to the ground, competing weeds all around.

The patch of evening primroses I visited is affectionately known to biologists as "The Pit." The Pit had been mined for its sand earlier in the century. From its bottommost point, I sensed something threatening looming over every horizon.

For more than a century, the Antioch Dunes have been severely degraded by sand mining, invaded by exotic grasses, encroached upon by a wallboard factory, smothered by gypsum dust, trampled by recreational visitors, and, formerly, sprayed by insecticide use on adjacent vineyards. The size of the naturally vegetated dune field has shrunk from over 200 acres to less than 30. A few thousand evening primroses remain on less than 12 of those acres, existing on their own nowhere else in the world.

Even after these few acres were nominally protected as part of Antioch Dunes National Wildlife Refuge, you could sense they still faced troubles of all sorts. The evening primrose and wallflowers were once trampled by "environmentalists" who tried to save Humphrey the Humpback Whale, a celebrated cetacean that was stranded in the river channel adjacent to the dunes in 1987. There is a touch of black humor in his voice when Bruce Pavlik talks about Humphrey "The Killer Whale." Pavlik notes that it is far easier for the public to attend to a huge whale found out of place than to a small patch of plants still in place but endangered by a society even further off any rational course.

When Pavlik and his students attempted self-pollination of this evening primrose at a garden outside his office at Mills College, in nearby Oakland, they confirmed what their field studies had already suggested: it was an obligate outcrosser—that is, unable to develop ripe seeds without the movement of pollen between plants. At the Antioch Dunes, far less than half the ovules on this wildflower matured into ripened seeds during a two-year period of drought when it was intensively studied in its remaining habitat. In 1987, a mere 20 percent of evening primrose ovules developed into mature seed, only a third of the percentage achieved by more common evening primroses in central California that year. Of the total number of evening primrose ovules produced in the Antioch Dunes in 1987, Pavlik estimates that 65 percent of them failed to mature into seeds due to pollination limitation.

Pavlik admits how disappointed he was by the paucity of insect appearances he and his students observed around the evening primroses: "We stayed out night after night, hoping to catch a hawkmoth, and never noticed a single one of them." Recent workers have yet to see an adult hawkmoth working the flowers there, although a sole hawkmoth larva was seen on the stem of an evening primrose in 1994. Like most evening primroses in California, the Antioch Dunes subspecies exhibits all the floral adaptations necessary to attract hawkmoths from dusk

*On a few acres of fragmented sand dunes in Antioch, California, live a few hundred individuals of a rare evening primrose subspecies* (Oenothera deltoides howelli). *Its pollinators—sphinx moths in the genera* Hyles and Manduca—*have not been seen at this locality for decades due to large-scale regional use of pesticides and herbicides.*

through early evening. Apparently, the adult hawkmoths do not remain there. Entomologists have suggested that hawkmoths have been scarce as floral visitors to the Antioch Dunes for at least 35 years. They may have been depleted during the days when pesticides were in heavy use on the cultivated grapevines nestled into the dunes. Those vineyards are now abandoned, but the moths have not resurfaced.

In the absence of moths, Pavlik's students recorded some bumblebees, a rather small *Sphecodogastra bee,* and a few flies visiting the flowers. The bumblebees appeared to be the only effective pollinators among them, for large amounts of pollen stuck to their abdomens before they moved onto the next plant, where they often brushed the stigmas of another evening primrose flower. Although bumblebees are not abundant at Antioch Dunes and they were only seen visiting the flowers for a brief period a couple hours after dawn, they may be responsible for the little seed set that does occur in the wildlife refuge. After the drought of the late 1980s, seed set has increased—perhaps due to visitation by a variety of generalist insects surviving among the plants. There are still no signs that the hawkmoths are making a comeback.

Pavlik continues to worry about the fate of the Antioch Dunes evening primrose. For even though it is capable of high seed production per plant, it has been constrained by the paucity of effective pollinators for a number of decades. Furthermore, its "seed bank" in the sand is meager compared to other wildflowers on site that Pavlik has studied. The Antioch Dunes evening primrose "has a smaller residual bank and is therefore less buffered against the erosion of genetic variation and against catastrophic events that reduce the adult population."

Genetic erosion, as Pavlik points out, is inevitably occurring. If we look at the diminished biotic interactions in a habitat fragment containing a rare plant population, we can see that many factors will contribute to this erosion. Geneticists J. M. Olesen and Subodh Jain have outlined some vicious synergies that may accelerate whenever fragmentation disrupts interactions between plants and animals. In a small plant population, there will be a higher probability of *genetic drift,* for example, and the loss of genes through this means. And because there are fewer potential mates, inbreeding is also likely to reduce *heterozygosity,* a measure of the population's genetic diversity. An inbreeding depression would lead to both lower seed production and lower seed germination rates—as Eric Menges has documented in small plant populations of prairie mustards from the Midwest. Once seedling recruitment no longer keeps pace with adult plant mortality, the population size may become so reduced that it no longer attracts a sizable fleet of pollinators, as has occurred with *Banksia* in Western Australia. Reduced gene flow will be evident among the plants that are pollinated, but many will go unpollinated and therefore fail to set seeds at all. Peter Leseica has demonstrated that when pollinators are lost or excluded from a rare Montana mustard, not only does the seed set drop but the seeds that *are* set are highly inbred and their seedlings have reduced fitness. At this point, seed dispersers may have so few rewards available that they abandon the site, resulting in reduced seedling recruitment, and in even less gene flow between plants.

Once pollinators and seed dispersers abandon the site, other plant species that they formerly serviced may begin to decline as well. As Beverly Rathcke and Eric Jules have warned:

> The pollination success of one plant species can also be influenced directly by the presence of other plant species that maintain pollinators. . . . [For example], a failure in flowering of an early flowering species caused migrating hummingbirds to leave the area. As a result, a later-flowering species experienced lower visitation and lower seed set. The disruption of such sequential mutualisms could cause cascading extinctions through the community.

Such a warning about the ultimate effects of habitat fragmentation must be placed in context. If habitat fragmentation by physical destruction of vegetation and chemical elimination of biota is not pervasive, then the case studies cited here may be the exceptions, not the rule. But what if they are just the tip of the iceberg—the most glaring examples of disrupted plant/pollinator relationships—while many similar cases have escaped our notice, below the scan of our collective vision, out of sight, out of mind? What if habitat fragmentation has begun to threaten tropical plants and their pollinators as much as those in the desert and temperate zones? What if our estimates of rates of habitat destruction fail to take into account the rates of biotic depletion in standing vegetation due to pesticides, hunting, and other pressures?

The story is multifaceted, but one thing is clear: there are fewer pollinators in the cold icy seas around habitat fragments than there are on "the visible tip of the iceberg."

*Near the small village of Sierra El Campanario in the Sierra Madre mountains of the state of Michoacan, Mexico, are about ten mass aggregation or overwintering sites of the monarch butterfly* (Danaus plexippus), *whose populations originate in the United States and Canada thousands of miles away. Here they hang in gaudy festoons from the sacred oyamel firs except on sunny days when they take flight to sip nectar and take water.*

# CHAPTER 7

# Need Nectar, Will Travel

## Threats to Migratory Pollinators

G ARY REMEMBERS:

A butterfly wind. We were enveloped in a butterfly wind, but we could still hear the incantations of warblers, orioles, and grosbeaks, as well as the distant echoes of marimbas. Steve and I had made a pilgrimage with friends to the Mexican wintering grounds of the monarch, one of the most active pollinators and nectar-feeders on milkweeds the world has ever known. We were astonished to find ourselves surrounded by some 12 million butterflies concentrated within just 3 acres. Many of them fluttered in the cool February air above, while others shuddered by the thousands on each nearby trunk of oyamel fir. These "sacred firs" formed a broken forest covering the slope of the Sierra El Campanario at 8,700 feet in elevation.

Not all of the monarchs around us could still fly, though, or even muster a shiver. One weakened monarch had slowly ambled around the herbs between my feet, then collapsed on the ground. It had arrived in Michoacan three months earlier, after up to a 2,000-mile flight from the north, and any fat reserves it might have had left were now nearly exhausted. The long migration and the cool winter had so depleted its strength that it would be unable to gain enough calories from floral nectar here to survive even another week. As many as 20,000 monarchs per acre may fall to the ground during the average overwintering season at the El Rosario sanctuary, prey for grosbeaks, orioles, mice, and ants.

I counted three dozen dead and dying butterflies within inches of our guide's feet, forming an orange, black, yellow, and beige butterfly mulch over the volcanic cobbles of the mountainside. An inchworm looped haltingly among this invertebrate carnage, its colors nearly the same as the now lifeless monarchs. Others remained among the living: one came to rest on Steve's head, while another attached itself to my jeans, where it shivered for awhile before returning to the quiescent masses clinging to the closest oyamel fir.

Right above us was another sight to behold. Flamboyant shingles of tens of thousands of monarchs swayed in the breeze, blending in with the boughs of the sacred fir as if they were fir needle clusters. I craned my neck back and looked straight up: there were thousands of other monarchs in each column of air that filled the spaces between the trees, for they were aloft as soon as the first beams of sunlight penetrated the forest, warming them and stimulating their flight to nearby nectar plants and pools of water to slake their collective thirst.

When I spoke with lepidopterist William Calvert, who had spent much of the winter attempting to census the monarch overwintering grounds of the Sierra Madre in Michoacan, he estimated that up to 30 million butterflies were concentrated in little more than 12 acres scattered over just five officially protected sanctuaries in the adjacent mountain ranges. Another of our traveling companions was Bob Pyle, founder of the Xerces Society for invertebrate conservation. He called it "the greatest aggregation of a living organism—and its deceased kin—anywhere in the world." An elderly Mexican peasant put it another way as he walked with me along the trail at El Rosario: "This is the monarch's sanctuary, but also their pantheon . . ."

The old man was right. Tens of millions of monarchs survive the winter thanks to the peculiar microhabitat offered by Sierra El Campanario, but millions may die there as well. Over most winters, only 5 to 10 percent of the entire population is killed by the slicing beaks of orioles, the chomping bites of grosbeaks, the nibbles of some 40 other potential animal predators, or the chilling frosts of winter nights. But our local guide, Homero, told me tales of what happened to monarchs when blizzards hit the highest ranges in Michoacan. When one freezing snow arrived a few years ago, a third of all the monarchs were killed in one fell swoop.

Vulnerability to this kind of unpredictable event may seem at first glance to be par for the course—migratory butterflies must take their lumps with everyone else. The trouble is that migratory monarchs now aggregate and overwinter at so few sites because much of the formerly suitable habitat for them has been irrevocably altered. The monarchs are very finicky about forest canopy structure. In fact, they settle in only a few of the possible forest locations available to them in their wintering ranges in coastal California and the Transverse Neo-Volcanic Belt in central Mexico. Although these two regions have completely different sets of trees in their forests, both offer closed forest canopies providing a narrow range of direct solar radiation and reflected light, offering most monarchs a buffer from climatic extremes. The forest stands are also associated with winter nectar sources, enough standing water to serve thirsty monarchs, and cool moist air pools that keep them from further dehydration.

Intact forests of sacred oyamel fir now make up less than 2 percent of Mexico's total land area—and within that 2 percent, sizable forest patches with the necessary water, nectar-producing plants, and suitable temperatures cover an even smaller area. Many once favorable forest habitats have already been cleared for farming, or logged, or overgrazed by cattle, or otherwise degraded by bark beetles and dwarf mistletoe. (These last two are small-scale natural occurrences.) With only ten overwintering roosts totaling less than 100 acres suitable for monarch use in central Mexico—and only five of them legally protected—migrating butterflies have perhaps begun to aggregate in concentrations much higher than their historic sizes.

The same trend appears to be true for California's overwintering monarchs. More than 21 sites historically frequented by monarchs have been destroyed, and another 7 of the 15 remaining have been severely damaged by California land developers within the last few decades. Accordingly, Bob Pyle was successful in getting the World Conservation Union (IUCN) to decree that the monarch migrations are "threatened phenomena" even though the species itself, *Danaus plexippus,* is not globally endangered.

But there is another view of the large-scale monarch phenomenon. Biologist Richard Vane-Wright has argued that the highly aggregated overwintering sites of monarchs in California and Mexico were in fact *caused* by the deforestation of 300 million acres of North America, a process that began in the centuries following Columbus. According to Vane-Wright, it was only *after* 1864, when deforestation was well advanced, that the monarchs began to expand their range and concentrate in just a few overwintering sites. The opening of forests into meadows and pastures, he says, actually increased the range and abundance of the milkweeds that monarch larvae and adults rely on for the alkaloids. (They sequester bitter-tasting poisonous milkweed alkaloids in their body tissues to protect them from predators—who get a dose of "one-trial learning" after eating one of these beauties and then vomiting it up.) Vane-Wright contends that "spectacular annual migration cycles, found only in the populations of North America monarchs, have evolved in their present form as a result of the cataclysmic ecological changes wrought by European colonists—[what he calls] the Columbus Hypothesis."

Whatever the case—whether the migratory phenomena of monarchs are on the increase or are "endangered"—an incredibly high proportion of all monarchs spend a third of the year gathered at just a few sites. Other migratory species face similar problems—most rely on a relatively small amount of habitat for a critical segment of their migratory cycle. If that habitat is vulnerable to development, or even to extreme fluctuations in weather, the migrants are like so many eggs in the proverbial basket.

But migrants such as monarchs are vulnerable in another sense as well because they have so far to go—so much territory to cross, so much time in which something may go wrong. When you consider the against-the-odds success of a long-distance migration lottery plus the need to pass through areas while their flowers are still providing nectar, it is amazing that migrating pollinators do not suffer more frequent catastrophic declines from natural disasters alone.

This is one of our main concerns with regard to three migratory nectar-feeding bats entering our U.S./Mexico border region, but it is just as apparent for certain monarch and other migrant populations as well. Consider for a moment the perils that migrants like two kinds of long-nosed bats must face as they move to their summer birthing grounds in the desert plains of New Mexico after wintering in central Mexico (and vice versa). These bats are not the only pollinators of the plants they visit, but even when other animals such as morning-visiting bees are present, seed set declines in years when the bats fail to arrive at these northern migration corridor sites. Although their numbers may not be in a steep unidirectional decline as once thought, they continue to face many natural and human-wrought challenges as they travel some 2,000 to 4,000 miles each year. This bat remains listed as an endangered species in the United States and is a major concern of Mexican conservationists involved in the Programa para la Conservacion de Murcielagos Migratorios de Mexico y Estados Unidos de Norte America, a binational initiative on migratory bats. To obtain enough energy to fly 100 miles a night and reproduce, these bats must follow what ecologist Ted Fleming calls "the nectar corridor"—a loop as long as 3,200 miles that follows the sequential flowering of at least 16 flowering plant species, including tree morning glories, several century plants, and giant columnar cacti. The bats that migrate northward along the corridor, sometimes as far as Tucson, Arizona, and Big Bend, Texas, must take their newborn young and then travel a safe route laden with nectar and pollen back through the U.S. Southwest and northern Mexico. Nowadays this transit is no easy feat.

Pretend you are a bat—or a monarch butterfly, for that matter—on your way south from the southern mountains of arid New Mexico in late summer. Your first peril may be an American rancher turned mesquite-killer, for many ranchers have sprayed their shrub-covered ranges with highly toxic herbicides such as 2,4-D, 2,4,5-T, teburithion, dicamba, or picloram to make room for exotic pasture grasses like lovegrass imported from Africa to feed American beef cattle. These chemicals may also kill your nectar plants. (They may also cause reproductive impairments not only in livestock but in wildlife or in family pets: increased postimplantation mortality, reduced birth weights, reduced growth, and reduced survival have all been recently documented effects from these rangeland herbicides.)

If you escape unscathed, you can cross the border into northern Mexico. There you will enter the Sonoran municipalities where the Border-Right-to-Know Project has recently documented rampant insecticide abuse on numerous small farms raising produce for export to the United States and elsewhere. Arizona Toxics Information Director Michael Gregory has determined that thousands of pounds of highly toxic insecticides are being applied to lettuce, cilantro, squash, and chiles in these northern Sonoran farmlands. A number of these chemicals, including paraquat and methomyl, are known to produce endocrine-disrupting effects on reproduction for several generations, not only in humans, but in many wildlife species as well. The bats, bees, and butterflies passing through these narrow valleys might not feed on the weeds or crops in these fields, but they would be susceptible to any aerially applied sprays drifting onto adjacent wildlands vegetation. Sadly, only one in 48 Mexican farmworkers interviewed in 1994 by Gregory and colleagues had any training from their government or industry representatives regarding which insecticides to apply to each crop at the appropriate stage. Moreover, pesticides not yet registered in Mexico were being purchased in the United States and brought across the border for use on crops for which they were never intended. And their deadly drift extends far beyond the crop rows themselves, affecting target as well as nontarget species.

If you make it past that point, you may again be exposed to herbicides used in the conversion of desert or thornscrub to nonnative pasturelands. Sonoran ranchers have already converted at least 930,000 acres of desert and subtropical vegetation in the very habitats where bats and butterflies must constantly find suitable nectar sources along their migratory routes. Ranchers sometimes combine the bulldozing of desert nitrogen-fixing legume trees and cacti with herbicidal spraying of shrubs and annuals before planting buffel grasses now favored as forage by cattle. Once planted, these grasses not only outcompete natives but carry wildfires into adjacent natural vegetation that is not well adapted to frequent burning. Cacti and century plants are particularly susceptible to fires carried by these exotic grasses. Nectar sources for moths and butterflies also decline where buffel grass has been introduced, for the grass cover suppresses seedling growth of natives.

In Sonora's irrigated valleys, you are likely to hit Mexico's Cotton Belt. There cotton is grown extensively—and has historically used at least five to ten times the amount of pesticides typically applied to food crops. In the central Sonoran river valleys, nectar-feeding bats have been found to contain DDT levels in their flesh four times higher (4.5

parts per million) than the historically high levels found in Arizona dairy cattle at the height of DDT usage in the United States. Nectar-feeding bats are not believed to "bioaccumulate" pesticides as much as the insectivorous bats that eat higher on the food chain. Still, they remain at risk wherever such chemicals happen to contaminate their food and water sources. Though surely not as vulnerable as their insect-eating cousins, most long-nosed and long-tongued bats have not yet had the luxury of returning to a pesticide-free diet.

Once out of the industrialized agricultural valleys and into the rugged mountain outliers of the Sierra Madre, you will be affected by thirsts and desires in subtle but pervasive ways. Bootleggers produce an alcoholic beverage called mescal bacahora from wild century plants. These agave species used to be harvested in ways that allowed for vegetative regeneration and periodic flowering, so small-scale traditional gathering did not threaten the nectar corridor. The old Mexican Indian traditions of managing wild century plants are fast disappearing, however, so that a greater percentage of century plants are having their growth terminated by unskilled harvesters. At best, the plants persist only through vegetative propagation and are never allowed to flower while they are in reach of an able-bodied harvester.

As a bat, however, you have a strong seasonal dependence on agaves along certain stretches of your nectar corridor. In a few spots, the corridor's continuity may be broken by overharvesting of agaves in southern Sonora. Over 1,200,000 wild agaves may be harvested each year in the state of Sonora alone—enough to diminish the availability of nectar to large flocks of migrating bats, and to reduce the agaves' genetic diversity. Old-timers in several Sonoran villages claim that overzealous harvesters and droughts have locally depleted agave populations that formerly were abundant. Today a person cannot walk a day's distance out of certain pueblos and collect enough mature agaves to make a gallon of mescal.

Farther back in the hills, from central Sonora through Sinaloa, you face another danger. Another kind of poison is affecting moths and butterflies, and perhaps bats as well. In fact, the spraying of paraquat on marijuana fields has eliminated Lepidoptera from entire valleys where they once were abundant. According to R. S. Peigler, a saturniid moth specialist, Yaqui and Mayo Indians have recently complained that they cannot collect enough silk moth cocoons to make the rattles that have been part of their ceremonial traditions for centuries. Apparently, paraquat spraying has drastically diminished the number of many moth cocoons being left in subtropical thornscrub vegetation.

And then you face dynamite. From Sinaloa southward, vampire bats have been preying on the blood of livestock for many decades. Recently, however, vampires have increased their range northward, coming within 150 miles of the U.S./Mexico border at Rayon, Sonora. In many places where vampires occur within ranching country, cattlemen seeking to eliminate vampires damage bat roosts in caves and rock shelters used by other species. Bat Conservation International has recently reported that Mexican free-tail bats and lesser long-nosed bats have been dramatically affected by the dynamiting or burning of bat roosts by cattle ranchers. Of the ten most important overwintering caves for free-tail bats in Mexico, over half have lost 95 to 100 percent of their populations. Such habitat losses have occurred at many sites throughout the ranges of nectar-feeding bats and butterflies in Mexico and the United States.

GARY REMEMBERS:
I once had a moment of recognition about the plight of migrant pollinators. Even when their roosts are protected in one area, the cumulative effects of disruptions along their entire nectar corridor may still diminish the numbers of arriving survivors. I witnessed this predicament firsthand one May evening when assisting with a nectar-feeding bat monitoring program in a borderland national park. I had volunteered to crawl into a mineshaft where up to 10,000 bats were known to congregate. Early in the warm season, perhaps only 5,000 had already arrived, and most of them were gravid females.

Arriving at the mineshaft entrance around eight in the evening, I sat with a U.S. National Park Service employee and watched as dozens of bats whirred by on their way out to feed that night on the nectar and pollen from cactus flowers. Twenty to sixty lesser long-nosed bats rushed out of the roost at a constant clip. They did not form a continuous stream as from some of the giant tourist bat caves; it was more like a froth let loose from a bottleneck. They took off in several directions, toward large stands of saguaro cacti within a 60-mile reach.

*A small group of three lesser long-nosed bats* (Leptonycteris curasoae) *hang from a dark cave ceiling roost after returning from a night on the Sonoran Desert. Their floral liaisons are betrayed by their whiskered muzzles splattered with pollen from century plants in the genus* Agave.

After a half hour had passed, my friend and I put on our headlamps and began the slow crawl through the horizontal mineshaft toward the monitoring equipment. Guano stench bit at my nostrils as we moved along, spotting as we went the serpentine trackways of rattlesnakes and paths where desert tortoises had entered the shaft in the past. All the way, I could hear loud whirring and shuffling sounds, not unlike rain and wind on the roof in a storm.

Just before reaching the monitoring equipment, I noticed something grim: recently fallen bats were lying dead, in the guano, being consumed by dermestid beetle larvae. In just one section of the mineshaft near the monitoring equipment, I tallied between 50 and 100 bat skeletons, intact or disarticulated, with leathery hides stretched over them. I wondered how many more lay beneath the roost, dozens of yards further into the mountain, but I didn't crawl in to look—some of the females may have been giving birth in there, and did not need to be disturbed.

What caused the death of so many bats within just a few days? This question remains unanswered for me. Did already weakened bats attempt to cluster around the warmth and hum of our temperature and humidity probes, our technical data logger and battery? Did exposure to pesticides in Mexico finally catch up with some of the bats at the northern limits of their migration? Had they arrived out of synch with the local flowers, due to their hurrying through an area of desert that had been deforested, converted, or destroyed? Whatever the answer, the corpses of bats were now afloat in a sea of guano and dermestids.

Like the monarchs, the lesser long-nosed bats are not as rare as most truly endangered species. A few fallen butterflies or bats do not mean that their kind is globally threatened. What is strikingly similar about monarchs and nectar-feeding bats is that each of these species aggregates into so few populations for a good part of each year. There are 33 threatened species of Mexican bats that roost in caves, but according to Mexican biologist Hector Arita, the lesser long-nosed bat is one of only two that nests in colonies of greater than 200 individuals. As for butterflies, the five monarch roosting sites in Michoacan collectively contain 20 to 50 times the number of monarchs in all the winter roosting sites in California combined. If one roost is destroyed, a fiftieth or a twentieth or perhaps even a tenth of all living individuals in the species may vanish with it in one moment.

Some plants—such as certain agaves—have devised a means of surviving such sudden fluctuations in pollinator numbers. Agaves with umbrella-shaped inflorescence may have been originally shaped by bat visitation behavior, yet their flowers remain generalized enough that bees and even hummingbirds will transfer pollen from one flower stalk to the next. Even when other pollinators are scarce, some agaves have another fallback strategy. Once the unvisited, unfertilized flowers wither, they produce small plantlets called *bulbils* in their stead. These bulbils are essentially parasitic on the mother plant and genetically identical to mom—they lack the genetic diversity associated with sexual recombination. But they allow mom's genetic legacy to persist until a pollinator returns to enable outcrossing.

Not all plants take out such life insurance policies. If they are to any extent reliant on migratory pollinators, their seed-setting abilities are susceptible to any and all fluctuations in pollinator numbers, whether due to natural or human causes. The plant's vulnerability increases with the length of the pollinator's migratory route, with the degree of disruption of nectar sources along the way, and with the intensity of aggregation of the pollinator's populations. Whenever too many eggs are put in the same basket, or whenever the basket has traveled too far over ground too dry or too rough, the results are likely to be broken, scrambled, rotten, or parched.

If migratory bats or monarchs were the only ones to deal with the perils found along the nectar trail, this story perhaps would be unremarkable. But throw in other kinds of nectar-feeders: the thirteen migratory hummingbirds, three sapsuckers, two warblers, and five orioles that move between the tropical and arctic reaches of the New World. Then consider the flying foxes that move between one island and the next in the Pacific. Then toss into the ring the hawkmoths that have been found to move between mountain ranges on successive nights. While most plants remain sessile, permanently rooted, there is a whole fleet of animals out there risking their lives as they serve as connective tissue between pollen donors and receptive stigmas for the local plant communities. Whether anther and stigma are a few feet apart or a few miles from one another, their animal intermediaries increasingly find that it is indeed a jungle out there.

*A groggy male squash and gourd bee* (Xenoglossa angustior) *has just climbed out of its noctural roost in a zucchini squash blossom. Soon it will begin a daily quest in search of female bees of its species as they visit blossoms for nectar and pollen. Like many territorial male solitary bees, it is an important pollinator.*

# CHAPTER 8

# Holding the Globe in Our Hands

### The Relentless Pressures on Plants and Pollinators

*G*ARY REMEMBERS:

We passed the somewhat soggy, punctured, hard-shelled gourd back and forth between our hands, as if it were a miracle among us. For the first time in six years, the Okeechobee gourd had been found, still surviving somehow, in the Southeastern United States. With the help of an old friend, I had at last determined that *Cucurbita okeechobeensis* (subspecies *okeechobeensis*) was not yet extinct, though it was undoubtedly endangered within its wetland home.

Other naturalists had already given up hope that it would ever be seen in the wilds of Florida again, for it faced fierce threats on all fronts. The gourds were surrounded by exotic weeds—from moonvine to smartweed—that handily outcompeted most plants indigenous to this swamp of custard apples. The custard apple trellises, which had for decades served the climbing gourd vines, were now in decline as the Southern Florida Water Management District played god with the water levels in Lake Okeechobee, sequentially starving, then drowning out, the remaining custard apple trees. Nearby, former habitat had been drained and planted with sugarcane. Where mosquitoes bred in the pools of irrigation tailwaters, aerial spraying kept both pests and beneficial insects in check. Even in "protected" wildlife areas in southern Florida, such spraying has been found to affect nontarget insects for distances up to 750 yards.

Long after I stood waist-deep in those alligator-infested swamp waters, I realized there was one beneficial insect that should have been buzzing around that custard swamp but has never been found there. Among entomologists, this big orange and brown native bee goes by the name of *Xenoglossa strenua*. This solitary bee has been found nearly everywhere squashes, pumpkins, and gourds have been grown in the southern United States and adjacent subtropical Mexico. If the gourd had been there for centuries on the shores of Lake Okeechobee, as I suspect it has, then this big mahogany-colored bee should have been there as well. In all the Americas where gourd blossoms have flowered since prehistoric times, there is simply no place that lacks this bee or its close relatives. In fact, squash and gourd bees are found in abundance on the Florida gourd's closest kin, the subspecies known as the Martinez gourd, wherever it grows on the Caribbean coast of Mexico. Why not here?

It has been ten years since the Okeechobee gourd was rediscovered on an island in the lake, but the full puzzle has not yet been put together. Recently, though, entomologist Mark Minno of Gainesville, Florida, rediscovered a second population of Okeechobee gourds within a mile or two of where naturalist John Bartram first described this species in 1774. There, in swamps along the St. John's River near Lake Dexter, Florida, Bartram observed "a species of *Cucurbita* which spread and ran over bushes & trees 20 or 30 yards high [and] which reflects on the still surface of the River a very rich and Gay picture." Mark Minno was much more matter-of-fact about his own discovery: "We just happened to be out there one day, and happened to see these gourds out in the swamps, and thought they might belong to the Okeechobee gourd species. We were just astounded that no one had really looked for it there since Bartram's time."

Minno did not have the luxury of floating a river as placid as that which Bartram describes: he found the gourd in high floodwaters while wading up to his armpits in another swamp inhabited by alligators. While his training in entomology allowed him to spot some melonworms—larvae of pyralid moths—feeding on the vines, he has yet to see solitary bees among the vines, despite his periodic return to monitor the gourds. There is fruit set due to honeybee pollination. Other biologists now intensively working with the Lake Okeechobee population of the gourds also report fruit set, but they too have yet to report any of the native bees.

A lack of reports, in and of itself, does not necessarily mean that the squash and gourd bee was once there but is now extinct. It is symptomatic, though, of a larger problem—that of missing information relating to the animal associates of some of the rarest plants in the world. As Peter Kevan of Guelph University in Ontario once pointed out, the information available on pollinators' interactions with plants is often the weakest link in our chain of understanding how ecosystems function.

In 1989, the Okeechobee gourd was listed among some 250 plants that the Center for Plant Conservation predicted would go extinct in the wild within ten years. Even if it continues to be grown in botanical gardens, there is a high likelihood that it will become functionally extinct in its natural habitat. It is without the animals that have been associated with wild gourds over the millennia—merely a "living Latin binomial" as ecologists Robert May and Anne Marie Lyles have called captive-bred species that are deprived of their original ecological contexts.

Filling such an ecological void can be difficult, as William Stolzenburg of The Nature Conservancy has documented for two Hawaiian species of *Brighamia*. The two rosette-forming plants are collectively known from barely more than 120 individuals left in the wilds of Molokai and Kauai. They rarely set seed any more unless they are hand-pollinated by what Stolzenburg describes as the "death-defying acrobatics of human moths." The human moths, in this case, are biologists from the Hawaiian Plant Conservation Center who rappel over the brink of 3,000-foot precipices above the sea. Then, hanging from

*Atop the Haleakala volcano in the Hawaiian Islands grows the "ahinahina" silver-sword* (Argyroxiphium sandwicense)*—an example of gigantism among a group of plants we normally pull out of our lawns as dandelions. These plants are now threat-ened and can no longer depend on a stable pollinator population, or enough nearby plants, for their reproduction.*

ropes, they brush precollected *Brighamia* pollen onto the stigmas of the few trumpet-shaped flowers growing out from the world's tallest sea cliffs.

These biologist-acrobats are trying to play the role once played by the native pollinators that *Brighamia rockii* and *B. insignis* have lost over the last two centuries. Today, only half of Hawaii's original set of nectar specialists persist—and many of the remaining nectar-feeding birds, from the Hawaiian *alala* crow to the crested *akohekohe* honeycreeper, are officially listed as threatened or endangered. Stolzenberg reminds us that the idea of *linked extinctions* suggests that one species' demise is triggered by that of its evolutionary partners. "If ever there might have been a stage set for exhibiting linked extinctions," he concludes, "Hawaii was it."

Stolzenberg may be correct in bestowing this dubious distinction on the Hawaiian Islands, but others might argue that Madagascar is just as vulnerable to linked extinctions. The Malagasy Republic on the island of Madagascar ranks among the top five countries in the world in terms of *endemism*—the percentage of its total species found there and nowhere else. Roughly 54 percent of its butterflies, 95 percent of its reptiles, 46 percent of its birds, 41 percent of its bats, all of its primates, and all of its thousand or so orchids are unique to the island. By the 1990s only 10 percent of Madagascar's original forest cover remained intact and less than 2 percent of its land base had achieved nominal protection. Deforestation, agricultural conversion, hunting, and overcollecting in the Malagasy Republic have depleted populations of one animal after another. According to the IUCN, 18 primates, 28 birds, 110 reptiles, 22 amphibians, 15 fish, and 45 butterflies native to Madagascar have suffered severe declines in recent decades.

If University of California conservation biologist Harold Koopowitz is correct, even these dismal numbers underestimate the severe loss of biological diversity from Madagascar. Koopowitz has developed a model that predicts the number of species extinctions. His model is based on the probability that random deforestation of various localities affects plant species known only from a few sites more than it affects widespread species. Unfortunately, 87 percent of Madagascar's well-studied orchid flora are restricted to three sites or fewer. Using an extremely conservative estimate that two-thirds of the island's forest cover has already been lost or damaged past its capacity to support orchids, Koopowitz predicts that over 500 orchid species must already have disappeared from Madagascar. If deforestation has left only 10 percent of Madagascar's natural vegetation sufficiently intact to support viable populations of orchids, the minimum loss already suffered is closer to 747 species. Koopowitz warns that these estimates allow "orchid survival" to be defined in the most minimal terms—that is, as single plants still persisting in a habitat. But many orchids are obligate outcrossers, and one individual is not enough to keep a population or species viable over any period of time. Koopowitz concedes that such minimally persisting species will probably die out, not because of random events, "but rather because of demographic problems that lead to inbreeding depression or inability to achieve pollination."

L. Anders Nillson and his Malagasy colleagues have provided on-the-ground verification of the trends the Koopowitz model predicts with regard to orchid pollination in forest fragments. Nillson focused on the *Cynorkis uniflora* orchid, which produces nectar in a floral spur that is accessible only to hawkmoths with very long tongues. In disrupted fragments of hilltop forests, the hawkmoth fauna available to orchids has become skewed—in fact, many more nectar thieves than legitimate long-tongued pollinators are represented. This imbalance has changed the degree of genetic variation in the remnant orchid populations, for long-tongued hawkmoth pollinators now draw upon relatively few plants for pollen. And the bulk of the orchid seeds produced are sired by a small number of pollen donors, indicating reduced genetic diversity. Nilsson is prophetic in his interpretation of these disruptions: "Interhabitat ecological links between larval host-plants, adult insects, nectar thieves and pollination are probably critical components in most pollination systems in the tropics. Destruction of forest habitats will inevitably cause more or less severe imbalances among pollinating guilds of animals across habitats, and interactions may become extinct prior to the organisms themselves."

In a more recent assessment, Koopowitz and his colleagues have broadened their concern from the fate of the orchids alone to their pollinators and seed dispersers: "In the real world, deforestation is probably hardly ever complete, and some individuals can be expected to survive the ax or the plow. Some plants can endure for extended periods of time under adverse conditions, but whether their pollinators and other commensals can also persist so that reproduction can continue appears unlikely."

Although the Koopowitz model was first applied to Madagascar, it has also been used to predict the fate of the neotropical floras of South and Central America. Five of the world's fourteen megadiversity countries—Brazil, Mexico, Colombia, Ecuador, and Peru—are located in this neotropical region. There 19 percent of the original forested area has been cleared since 1950. On the basis of locality data from 4,258 species in the neotropical flora, Koopowitz, Thornhill, and Andersen have predicted that 3,020 plant species formerly known from three sites or fewer have already been lost since midcentury. If current deforestation rates continue unabated, the entire neotropics may lose 70 to 95 additional plant species every year.

Let's see what the model suggests for just one country known for its megadiversity, Ecuador. Ecuador ranks in the top ten countries in the world in terms of species richness and endemism. One-eighth of its 1,120 butterflies are found only within its boundaries—giving Ecuador the third-richest lepidopteran fauna in the world. In addition, Ecuador may have been home to as many as 20,000 plant species earlier in this century. But Ecuador is also in the top ten countries in the world with regard to its rate of forest loss. At least 54 percent of the Ecuadorian forests standing in 1950 have been lost due to lumbering, agricultural conversion, and oil exploration. The Koopowitz model suggests that Ecuador may already have lost 3,275 plant species since 1950, or 16 percent of its flora. And given current rates of the country's deforestation, another 67 flowering plants are annually being driven toward extinction.

Orchids comprise a significant portion of these losses. For pleurothallid (side-shooted) orchids, which have their center of diversity in Ecuador, 402 of the 3,405 known species may have already been driven to extinction by deforestation. In Ecuador and adjacent nations, deforestation is contributing to the loss of one *Masdevallia* orchid every year and one *Dracula* species every three years.

Koopowitz, Nillson, and others emphasize the plight of orchids for two reasons. First, orchids comprise perhaps one out of every ten flowering plants on the planet. Second, orchids are often epiphytic—they grow upon other plants—and thus are extremely vulnerable in areas where trees are felled. Deforestation since 1950 may have already wiped out 22 percent of the 25,000 orchids that have been described by botanists. Simply put: by the year 2000, about one-fourth of the world's largest plant family may be wiped out as the result of the last half century of deforestation and land conversion in the New World and Old World tropical forests.

Earlier we noted that most of the so-called orchid bees do not have strict one-on-one mutualistic relationships with orchid species.

Nevertheless, many of these bees will surely be affected by deforestation's toll on orchid diversity. Let us assume—along with euglossine bee expert Dave Roubik—that one-half to three-fourths of all euglossines in the neotropics visit orchids with some frequency. Overall, orchid bee species may number from as little as 5 to as many as 50 per site, with 15 to 30 active in any particular month. Based on their surveys in central Panama, Roubik and his colleague James Ackerman have recorded 38 euglossine bees frequently visiting 51 orchids. (Another 19 bees visit them rarely or never. As many as 24 orchids can be found at a single site.) Because 36 of the 38 bees were found at multiple sites, they would not necessarily be driven to extinction should a single site be destroyed—as 11 of the orchids themselves might be. But 29 of the bees visited only one orchid species and thus would be particularly vulnerable should that orchid die out at all sites.

Roubik and Ackerman do not necessarily see the coevolution of orchids and euglossine bees as being so equally weighted that a bee would be lost for every orchid lost:

> Orchids are clearly dependent upon euglossine bees for pollinator service, and much specialization may have occurred via adaptation to specific pollinator pools. [While there is] no evidence that euglossines [exclusively] are dependent on orchids for volatile chemicals . . . species that did not carry pollinia [from orchids] were not abundant. These data imply specialized dependency and perhaps indicate that bee species cannot become abundant if they lack orchid hosts.

If over a third (37.9 percent) of all orchids are known from just one locality, and over two-thirds (68.2 percent) are found in three localities or less, deforestation will take its toll on many geographically restricted orchids and the orchid bees dependent on them. If a bee has even a partial dependence on a particular orchid species, chances are slim that it will find other nearby sites where the orchid might still exist. Unless it can switch to another, more generalized orchid to obtain the fragrances, pollinia, or nectar it requires, the bee will surely be faced with extinction.

To make matters worse, wholesale habitat clearing is not the only threat to plants and pollinators in the world's megadiversity countries such as Ecuador and Madagascar. Although many of the 26,000 plants in the world that are threatened with extinction are vulnerable to chainsaws, bulldozers, and plows, their habitats can also be degraded or fragmented in more subtle ways. Forests may be degraded without being clear-cut and the results can be just as devastating. In the Lomas Barbudal dry forest of Costa Rica, biologists studied bee diversity in intact forests and then returned 15 years later to find that the same site had survived clear-cutting but not fragmentation. They demonstrated that solitary bee diversity there declined due to a variety of factors: the loss of fragrant oils required for reproduction by male bees, the loss of female nesting sites, the impact of fires on nesting success, and the loss of nectar sources critical to the development of young bees. The average number of solitary bees per sweep, on the same day of the year at the same locality, declined from 70 bees in 1975 to 37 in 1989.

Such dramatic short-term declines in species composition are surprising, because most models predict a 50 to 400-year time lag before habitat fragmentation results in extirpations or extinctions. And yet these same models warn us that eventually we may not simply lose rare species, but some of the more dominant ones as well: the species that exert major control over ecosystem functions, as well as the species that are the most efficient pollinators of the world's flowering plants.

We are disturbed by the implications of one such model, developed by David Tilman of the University of Minnesota, along with Robert May and others at Oxford University. Their model of habitat fragmentation attempts to predict the extinction debt our generation is passing along to other generations—and when this debt will come due in the future. They have verified that even a slight increase in habitat fragmentation threatens many species, particularly in places where a large portion of a habitat has already been fragmented. Their model compares the consequences of added habitat destruction in cases where 20 percent of a region's vegetation has been destroyed. If you increase the rate of habitat destruction in that region by just 1 percent, extinction rates then become eight times higher if 90 percent of the native plant cover is already gone than if only 20 percent of the protective plant cover has been cleared away.

And yet Tilman and his colleagues warn us there are large differences in how habitat fragmentation plays out in tropical versus temperate forests. In one of their model simulation scenarios, destruction of just one-third of a tropical rainforest triggered the extinction of 35 percent of the total species in the area. Most of these extinctions did not occur immediately, but took up to 400 simulated years after the habitat became dramatically fragmented. In contrast, the same magnitude of destruction eliminated only 5 percent of the total species in a temperate forest—but those extinctions occurred within 40 to 60 years after habitat fragmentation. In general, the smaller, more abundant species (such as insect pollinators) gradually declined in the modeled temperate forests, while large vertebrates and other rare species were much more quickly affected by habitat loss in the tropical rainforests.

Habitat fragmentation also increases pollinators' exposure to invasive competitors, to parasites and diseases, and to pesticides. The Pesticide Trust and other organizations have raised their concerns about the effects of uncontrolled pesticide and herbicide use in five of the fourteen countries richest in biological diversity: Ecuador, India, Brazil, Malaysia, and Mexico. People and animals in these countries must face the heightened exposure to herbicides and insecticides that often accompanies forest clearing and wetland drainage to open lands for farming and ranching. In Ecuador, for example, one-fifth of the 6,200 tons of pesticides imported in 1990 comprised chemicals that cause reproductive abnormalities in humans and other animals: paraquat, carbofuran, dichlorvos, endosulfan, methamidophos, methomyl, monocrotophos, and phosphamidon. Pesticide sales to Ecuadorian farmers have more than doubled since 1980. And yet, as one Ecuadorian agronomist (who asked to remain anonymous) has confirmed, pesticide applications are seldom controlled sufficiently to reduce exposure to humans, beneficial insects, or other animals: "Selling pesticides to farmers in the way it is done in Ecuador is like giving strychnine to people who don't know what it is."

Similar reports have recently come out of Mexico, which dramatically increased its pesticide imports following the North American Free Trade Agreement (NAFTA). According to one Mexican official, more than 165 million pounds of pesticides were used in the country in 1993—and in the first six months after NAFTA went into full effect, pesticide imports jumped another 50 percent. Mexico had already experienced a decade and a half of escalating sales of pesticides, however, from $199 million worth in 1980 to well over $560 million in 1990. Mexican farmworkers are now spraying a number of highly regulated or banned chemicals known to poison bees and other pollinators: aldrin, carbaryl, chlordane, DDT, dieldrin, endrin, heptachlor, malathion, and parathion. Twelve of the chemicals currently being used in northern

Mexico are known to disrupt the endocrine systems of animals, including many pollinators, no doubt resulting in long-term declines in reproductive success among some species. In terms of their lethal dose toxicity for bees, all insecticides are not created equal. DDT is moderately poisonous to honeybees and for solitary bees, while highly toxic killers include such formulations as malathion.

There is considerable irony in the fact that Mexico is importing such large quantities of hazardous chemicals from the United states and Canada as part of NAFTA—for Mexico is home to many more native pollinators than occur in either of the other two North American nations. Consider the butterfly as an indicator of pollinator diversity. Mexico is home to 471 species of butterflies, 46 of them endemics; Canada has less than a third that number, with 150 species, only 2 of them endemics; and the United States hosts 292 species, only 22 of them unique. Essentially, all of the approximately 160 genera of bees known from the North American continent can be found in Mexico.

Brazil is a country of perhaps even greater concern, for it ranks third in the world in plant and animal endemism and fourth among all countries in species richness. Although its rate of forest loss is ranked "only" sixth among the nations of the world, this conversion of forests is clearly being followed by the intensive use of agrochemicals that potentially affect many pollinators. Brazil is both the major user and the major producer of pesticides in Latin America. Its farmers spent $2 billion on them in 1990—half as much again as all other users in the region combined. Brazilian pesticide sales have nearly tripled since 1980, so that nearly 5 pounds of pesticides are annually used by each of the 23 million farmworkers in the country. These chemicals are used to control mosquitoes, cotton pests, agricultural weeds in newly opened fields within fragmented forests, and roadside weeds. One Brazilian toxicologist has estimated that 280,000 Brazilians—an incredible 2 percent of the total population—are poisoned by pesticides each year. If humans are exposed in such quantities, it is almost too frightening to imagine the intensity of exposure of target and nontarget invertebrates. The severity of pollinator declines resulting from this insidious connection between land clearing and deadly chemicals is not likely to be known for decades. But there is no doubt that the rainforest is being cut down equally effectively by both metal and "chemical chainsaws."

When James Przeslawski recently interviewed Amazonian orchid expert Callaway Dodson, he realized that Latin America is quickly coming to the end of the never-ending forest. In Dodson's words:

> The old saying in Ecuador is "Fat parrots mean skinny children." When it's thrown at you that way, what do you do? The last two decades of destruction have been the result of overpopulation in which the government has sought to take the pressures off the cities by giving forests to colonists. The end result has been deforestation at an almost unbelievable rate; you can't believe how fast it is vanishing. And it's not slowing down; it's accelerating as the next generation comes into their reproductive agespan. If the reproduction stays that high, then there appears to be no hope. The parrots have had it. They're on their way out and so are the orchids!

Massive forests are being reduced to piecemeal fragments on other continents as well. We may someday see on the large landmasses of this planet what we already see on islands: once-common plants, formerly capable of attracting a diverse assemblage of pollinators, becoming limited to only one generalist pollinating species, such as the European honeybee, so common in many secondary patchwork habitats.

If one plant has a prophetic message for us, it is the kapok tree, the Mayan tree of life. Within the intact forests found on various continents, naturalists have recorded that giant kapok trees are actively pollinated by insects, perching birds, opossums, primates, bats, and hovering birds. Within an island population of kapoks in Western Samoa, however, only one animal pollinator is available to move pollen from one tree to the next for this obligate outcrosser: the flying fox, *Pteropus tonganus*. Flying foxes are in decline throughout the South Pacific. Should this species become overhunted or find its habitat deforested, the only pollinator for the kapoks on these islands may disappear. Tom Elmqvist, Paul Cox, and colleagues have put it this way: "Flying foxes are critical pollinators for forest plants on isolated oceanic islands where the pollinator fauna is depauperate. . . . The loss of flying foxes would have serious consequences for the long term viability of isolated island ecosystems."

This finding echoes a concern raised by many biologists—from the Smithsonian's Tom Lovejoy to Harvard's Edward O. Wilson and *Outside's* David Quammen. The point is becoming clear: Oceanic islands are no longer the most isolated ecosystems on this planet; some forest fragments now show the very same syndrome. These fragments can be thought of as islands awash in a terrestrial sea now stripped of many keystone mutualists including pollinating animals. No doubt we will begin to see depauperate pollinator faunas evident in forest remnants, physically and chemically denuded habitat fragments, on each of the naturally vegetated continents. As underscored earlier, fewer "islands" these days are tropical paradises for pollinators, and nothing about forest, prairie, or desert patches in seas of degraded landscapes will remind us of a bygone once-pollinated and fruitful Eden.

*In the dense lowland rainforests of peninsular Malaysia, emerging from the canopy, are immense trees* (Koompassia excelsa) *that for decades at a time serve as safe retreats for dozens of gregarious colonies of a fiercely defensive giant bee* (Apis dorsata). *The wooden herringbone-style ladder is meticulously fashioned by families of honey hunters who lay claim to these trees and sustainably harvest honey year after year.*

# CHAPTER 9

# Keepers of the Flame

*Honey Hunters and Beekeepers from Ancient to Present Times*

S TEVE REMEMBERS:
Malaysia's rainforest at two o'clock in the morning was unlike anything I had ever seen. I was surrounded by a dense stand of thin, straight-trunked, smooth-barked trees known as dipterocarps—nearly a thousand different kinds of tree per acre. Every once in a while, a palm crown or massive legume canopy towered above them—as they do in the neotropical rainforests with which I was more familiar—but these paleotropical giants lacked many of the characteristic orchids and bromeliads that drape down from rainforest canopies in Central America, where I had worked earlier in my career.

Yet there was something stranger still about the rainforest on this particular night: an ancient honey-harvesting ritual had begun, one that would be accompanied by an incredible pyrotechnic display. I stood not far from an enormous tualang tree, waiting with old and new friends to witness an astonishing shower of embers raining down from smoldering torches held 90 feet above us. Somewhere high in the tualang canopy, a 70-year-old Malay honey hunter and his 16-year-old grandson were readying their gear to gather honey from giant Asian honeybee colonies. On the ground below them, three singers chanted ancient prayers integral to the tualang honey-hunting ritual.

I had come with a dozen other bee scientists to see one of the few giant tualang trees near Pedu Lake, in Kedah province of northern peninsular Malaysia near the border with Thailand. Of course, I was fairly well versed in the many features that distinguish the native Asian honeybee, *Apis dorsata*—a giant reaching an inch in length—from the domestic European honeybee, *Apis mellifera*. Nevertheless, no amount of reading and prepping by Malaysian colleagues could have prepared me for the ways these honey hunters still directed their contact with the giant bees through animistic rituals—to cajole, charm, then finally calm the bees in order to gain access to their glistening honeycombs.

Here, I sensed, humans had learned to face the ferocity of social bees with their own magic, as they have done for untold millennia. Here I could see how some of the most tenacious beliefs about the healing power of wild honey have been made manifest, undiluted, still alive in an ancient, giant hive filled with stories perhaps as old as our species' ability to weave tales.

I was guided by Professor Makhdzir Mardan, a honey-hunting specialist from the Universiti Pertanian Malaysia near Kuala Lumpur. Mardan had learned much of the Hindu and Islamic symbolism about bees during a decade of visiting dozens of honey hunters. Out of the bundle held in his memory, he selected one fable to tell those of us who had become the newest pilgrims to the dipterocarp forest and the giant tualang tree:

"Seems in ancient times, there was a Hindu handmaiden called *Hitam Manis,* "Dark Sweetness," for she was a dusky beauty. She fell in love with the reigning sultan's son, who returned her love. But they could not marry, for she was a commoner. She and her fellow maidens—called *Dayang*—were forced to flee the palace, for the furious ruler wanted to kill her. As she ran away, a metal spear pierced her heart. She and her friends were turned into bees and flew way."

The professor paused, and suggested that the group lie down on the open sloping ground, to look up into the canopy of the tualang tree, in anticipation of what would happen next in the story:

"One day, the prince—now engaged to a princess—noticed a honeycomb high up in a tree. He climbed the tree for it and discovered a sticky, sweet substance inside. He called down to his servants for a knife and a pail. When the pail was hauled down, they discovered to their horror that the prince's body was in it, all chopped up in pieces!

"A disembodied voice cried out that he had committed a sacrilege by using a metal knife in cutting the comb, for *Hitam Manis* herself had perished from a metal instrument.

"Later, a golden shower by the bees restored the prince to his entirety."

The professor paused again, this time with irony in his voice, for he was referring to the "golden shower" made by the bees during their mass defecations just after sunset. This same "golden shower" was once confused with the dreaded "yellow rain," a deadly form of real biochemical warfare that poisoned thousands of villagers during the Vietnam War. It is now clear, thanks to the work of Chet Mirocha and other scientists, that the older yellow rain is far from deadly: giant Asian bee droppings enrich the tropical soils with massive quantities of nitrogen, a nutrient often in short supply in rainforests. The "golden shower" was given restorative rather than destructive qualities in this Malaysian fable—restoring the body of a man who would have been left fragmented for failing to pay respect to *Hitam Manis*.

Professor Mardan concluded by reminding the group how this legend continues into modern times to guide honey hunters in the way they harvest the delicate beeswax combs: "To this day, no metal—only equipment of wood, hide, and cow bone—is used at all stages, in deference to the early anguish of *Hitam Manis*."

Despite the risky, rough-and-tumble lives the honey hunters lead, they always refer to the giant *Apis dorsata* with great tenderness, calling the bees *Hitam Manis*, a lover worthy of a prince. They also show their respect for the giant bees by referring to them only indirectly, through poetic nicknames like "Blooming Flowers" or "Fine Friends." They humbly refer to themselves as the *Dayang*, the handmaidens of *Hitam Manis*.

While I vainly tried to see the honey hunters up on their makeshift herringbone ladders in the tree's canopy, the *Dayang* working below filled the air with more incantations. The honey hunting is always done after the moon has gone down, or on moonless nights, when the bees do not fly from their nests. The *Dayang* and all other observers on the ground are forbidden to carry torches or even flashlights near the tree. Otherwise, the 30,000 to 70,000 *Apis dorsata* in each of the nests up above would make a beeline toward the lights, stinging everyone in sight.

One day a few weeks before, when one of Professor Mardan's assistants didn't heed the warnings, he was attacked by the bees from just one colony in this tree. Receiving well over 200 stings, he had to be carried a mile down a steep trail, then hospitalized, and did not return with us. We, like the *Dayang*, would do anything to avoid the wrath of the giant bees—considered the most pugnacious of the seven or so species of *Apis* native to the Old World. Compared to these giant bees, the Africanized "bravo bees" now found throughout much of the Americas are but minor irritations.

The group stood tensely below the 120-foot-tall tualang—a tree legume known to science as *Koompasia excelsa*. The two Malay men, we realized, were risking their lives climbing 90 feet into the canopy on sturdy but flimsy-looking herringbone ladders. The older man, Pak Teh, had been climbing this very tree since 1965, but still adhered to all the precautions and taboos. The wiry 70-year-old had ritually bathed, prayed, then left honey offerings at the base of the tree before beginning his ascent around one in the morning. His grandson had participated as a *Dayang* watchman on the ground before, but this was the very first time the boy would make the climb to serve as honey hunter and bearer of the gigantic torch they would use.

By three in the morning, they were so high in the tualang's canopy that I could not have seen them even if there had been light. They had purposely waited until the last vestiges of moonlight had vanished below the horizon, then signaled the *Dayang* to send up a huge cowhide bucket on the rope-and-pulley hoist they had constructed. One of the *Dayang* "handmaidens" reinitiated the singsong incantations that would rise sporadically through the rest of the night, until nearly daybreak. We could just make out the faint silhouette of the giant tree, the three climbers, and the scattered bee combs against the brilliant zodiacal pinpoint lights, another type of fire in the sky.

Finally, I could hear Pak Teh calling down that he was clinging to an immense branch above the first bee nest. At last he was ready to take honey and brood from one of the eighty-some colonies suspended like oversize Christmas ornaments from the tree. All of the *Dayang* were now chanting loudly, hoping to appease the bees and ensure a safe harvest. Pak Teh asked for the smoldering torch made of a tightly bound tail of selected liana vines. The long torch had been ignited at the base of the tree as part of the preliminary ritual. Occasionally we could see its incandescent tip like the red glow of a cigarette in a darkened room, or catch a glimpse of a cascade of sparks when someone jostled it against the tree trunk. From their precarious position, Pak Teh, his grandson, and the third climber began to direct the torch's heat, sparks, and smoke toward the comb, arousing the bees from their slumber.

Then, beginning the "shower of sparks," the men directed the torch's burning tail against the massive parabolic comb, singeing it and brushing away its thick curtain of bees. Suddenly Pak Teh started to bang the smoldering end of the torch vigorously against the limb just above the giant comb. Glowing orange and reddish embers of all sizes rained down by the thousands like a phantasmagorical meteorite shower. No Fourth of July fireworks display was ever so memorable or long-lasting for me as this native pyrotechnic event that special night in the rainforest near Pedu Lake. Lazily, the cascading sparks fell in wavering trajectories toward the clearing below the giant tualang. Until then I had been craning my neck below the tree, but now I moved back into the protective cover of the dipterocarp forest.

As some of the glowing sparks reached the ground, I heard an ominous roar following them out of the tree. It was unlike anything I had ever heard. Tens of thousands of incensed bees were swarming out from their comb, following the rain of fire from the giant bee nests down to the earth. Bees from neighboring colonies took flight, too, and joined in the noisy melee. They were so loud, so very close, that I instinctively ducked, and held my breath—since the carbon dioxide in your breath is a stimulus that elicits stinging in bees. I felt shaken to the core by the unshakable notion that thousands of defensive bees might find *me*, along with the *Dayang*, huddled together in the darkness far below.

Instead, the giant bees dispersed, clustering around the places where the embers had vanished. Harmlessly they settled on the ground and the surrounding foliage, where they would remain until the first light of morning lured them upward again to the remnants of their despoiled nests. By that time, however, Pak Teh would be gone from the tree, and the honey from some of the colonies would be harvested.

While we still huddled together, stunned, Pak Teh had already started extracting honey from the nest high above us. Using the shoulder bone of a cow, the scapula, he began to cut away at the comb. He then folded a 3-foot-long segment of wax together, so that its sweet golden treasure was captured inside, a movable feast. This comb section he then draped over the cowhide bucket, which he filled until it overflowed. Each time he lowered the bucket of honey and comb to the men below, another cowhide and rattan bucket was hoisted up.

Pak Teh moved from one colony to the next over the following four or five hours. Finally, at six in the morning, the sinewy old man and his two helpers reached the ground exhausted. There the *Dayang* helped them nurse the few stings they had endured during the night. (They treated Pak Teh's welts with giant bee honey, for they considered it a sure cure!)

As Pak Teh and his grandson had filled bucket after bucket, the *Dayang* below would carry the comb fragments over to an immense boulder overhanging the entrance of a cave filled with nectar-feeding bats. There the men squeezed and filtered the remaining honey from the beautiful wax combs. Hungry, tired, we huddled around the messy harvest and joined the *Dayang* in a joyously sweet meal of honey and bee brood. Some of us fought for the best chunks of comb—with larvae and pupae—to plop them into waiting liquid nitrogen containers for DNA analysis back home. It was an incongruous contrast: on the one hand, the ancient honey harvest ritual; on the other, the modern technorace for the hidden genetic secrets of these fearsome giant pollinators. Meanwhile, under the boulder, the Malaysian men continued to filter the honey into 7-gallon vessels waiting to be transported back to nearby village markets.

All together, Pak Teh and his crew would gather nearly a thousand pounds of honey from this tree. It was but one of about ten trees they tend, trees widely scattered through the forests surrounding Pedu Lake. The Tualang honey hunting allows each man in the *Dayang* crew to make as much as $150 more per month than he would make as a rubber tapper or farm laborer. Today the hunters must obtain permits to harvest the honey trees from the Sultan of Kedah province, for competition has become as fierce as the bees themselves—there are now more than 70 crews of tualang honey hunters who compete annually for the few giant honey trees found in their region. Collectively, they have harvested as much as 150,000 pounds of honey in a single two or three month season. That yield may decline, however, as deforestation further impinges on their activities and, of course, the lives of the giant bees in their rainforest home.

Professor Mardan is outspoken about how logging in the nearby forests is now disrupting this ancient tradition: "To many honey gatherers, as long as there are bees and the forest, there will always be honey gathering. But now the biggest threat to the profession of honey gathering seems to be the depletion of floral resources in the tropical rainforest brought about by indiscriminate logging. Many professional honey gatherers feel strongly that bee trees—which make good timber—should be protected from loggers. Urgent attention should be paid to their opinions."

Some foresters do, in fact, concede that the giant tualangs may be among the tallest trees left on the face of the earth, reaching heights of 150 feet or greater. These trees hold the vestiges of one of the oldest relationships between bees, floral resources, and their human stewards. They also offer honey hunters a way to supplement their meager incomes in a sustainable way without depleting the forest—and without disturbing the giant bees to the extent that they abandon the trees forever. Pak Teh himself has watched bees return to the same tree he has worked every year since the mid-1960s. The ancient prayers are still offered—and he still uses a cow scapula instead of a metal knife and a wooden ladder instead of an aluminum one—in order to honor *Hitam Manis,* the dark sweet goddess for whom he and his family are handmaidens.

Far away across the Eurasian continent, in eastern Spain, ancient petroglyphs on a rock shelter wall depict an activity similar to the one witnessed near Pedu Lake. At Barranc Fondo near Castellon, your eyes climb the sheer rock wall where five human figures ascend a ladder toward a nest, while families wait below, eager for a taste of honey. The ladder reaches to a nest surrounded by bees, a nest high up in a tree. Spirits of grazing animals and other creatures, branched plants, and rounded cobbles—bats, birds, or more bees—float around the crown of the tree. The petroglyphs of wild honeybees at Barranc Fondo are estimated to be at least 6,000 years old. Similar images of ladders leading to nests of giant Asian bees can be found at Jambudwip Shelter at Pachmarhi, India, dating from 500 B.C.

Eva Crane, the founder and past director of the International Bee Research Association, has documented a variety of prehistoric rock paintings of honeybees reaching back to 10,000 B.C., perhaps earlier. Crane claims that the petroglyph legacy is one of the many suggestions that early human cultures "developed an affinity with the bees from which [they] took honey—in spite of the stings—and the bees were often revered as magical, or even divine." Environmental historian Keith Thomas reminds us that this human affinity with colonial pollinators was expressed in remarkable ways:

> Bees could be communicated with, for, when they swarmed, their owners would whistle, clap their hands, ring bells and tinkle basins and kettles. This was an ancient practice, going back to Roman times, but still universally observed in eighteenth-century England. Its original purpose seems to have been to warn neighbors of the approaching swarm and to prevent disputes by establishing one's ownership in advance. . . . But by early modern times the noise was widely regarded by country people as a means of addressing the bees themselves. It was thought to prevent them from flying very far; it made them "knit" and encouraged them to settle sooner.

Prehistoric petroglyphs of honeybees and associated rituals are not restricted to Europe. Thousands of African rock paintings depict ladders leading to honeycombs and caves, many of them surrounded by swarms of attacking and confused bees. From the curves in the petroglyphs symbolic of bee's nests in Algeria to honey hunting scenes in Zimbabwe, there is ample evidence that the collection of honey from little *Apis mellifera* was as ritually imbued as the collection of honey from giant *Apis dorsata* continues to be today.

*A petroglyph, or rock painting, from the Arana (Spider) Cave at Bicorp, near Valencia in Spain. A lone honey hunter climbs a rope ladder to harvest honey, beeswax, and probably brood from the comb of honeybees high on a cliff face.*

It is hard to pinpoint when *beekeeping* as opposed to honey hunting actually began. Eva Crane thinks the earliest unambiguous evidence of beekeeping—tending bees in an apiary in handmade containers—is depicted in four scenes from the temples along the Nile in Egypt. These painted panels range in age from 2400 to 600 B.C. The scenes apparently depict the harvesting, processing, and storing of honey in large clay vessels. In the tomb of Pabesa, painted and incised reliefs dated between 660 and 600 B.C. show dynamic associations between beekeepers and flying honeybees: their hives and honey storage vessels.

The beginnings of beekeeping, however, could date back long before that, well before hives fashioned from bottlegourds, pottery vessels, or wooden boxes came to be. The earliest tended hives may simply have been *boles* of already inhabited tree trunks, sawed loose, detached, and transported to human settlements where honey-loving bears, badgers, or skunks could not reach them. Centuries later, the Egyptians became the first migratory beekeepers, floating hive-laden barges up and down the Nile, providing pollination services to floodplain farmers while making a honey crop.

*By 3500 B.C., the Egyptians were already advanced beekeepers floating their hive-laden barges up and down the Nile—the first migratory beekeepers. The bee hieroglyph shown here was used to denote the king of Lower Egypt.*

Beekeeping can be dated through the artifacts associated with it, but when did humans begin to understand that bees and other creatures were responsible for pollinating plants? And when did humans themselves begin to pollinate crop plants to produce higher fruit yields? Archaeologists again turn to Egyptian temple hieroglyphs as documentation of the hand pollination of dates, interpreting scenes dating from 800 B.C. Noticing that female date palms found in isolation from pollen donors produced less fruit, perhaps date cultivators struck on the notion of transporting and brushing male inflorescences against those on fruit-bearing date palms. This practice may have begun among the Sumerians and Assyrians, subsequently spreading to other Mediterranean cultures.

While not all cultures around the world have practiced hand pollination of their most valuable crops, many have their own terms and symbols for natural cross-pollination. Among the Native American maize farmers of the Upper Missouri River floodplain on the Great Plains, there is an ancient parable reminding children that "corn can travel." The famed Hidatsa gardener known as Buffalobird-Woman knew that different strains of corn planted within "traveling distance" of one another would become cross-pollinated and thus contaminated or multicolored. Because each corn variety "had not all the same uses with us," her people would endeavor to keep each one pure by planting it a considerable distance from the others.

It would be wrong to think that indigenous farmers who have been taught hand-pollination techniques by outsiders were previously unsophisticated in their management of cross-pollinated crops. Such an implication underlies the story of how Europeans trained the Totonac Mayan Indians to pollinate vanilla, an orchid native to Mayan homelands in Vera Cruz, Mexico, that requires cross-pollination. Some claim that Mexican Indians were unaware that their native *tlixuchitl* flowers required cross-pollination to produce abundant vanilla pods. Not until 1836 did a Belgian botanist named Morren successfully hand-pollinate vanilla flowers maintained at a botanical garden in Liège. Word of the technique quickly spread between vanilla fanciers in Europe, many of whom adopted the techniques for their greenhouse plantings of orchids. Within a decade, French immigrants to Mexico began to use such hand-pollination techniques to obtain a fivefold increase in vanilla pod yields in the Nautla River basin of southern Mexico. Folklore has it that the French colonists were soon harvesting many more vanilla pods from their limited plantings than their Indian neighbors were getting from large plantings. The Totonacas accused them of stealing from their own plantations. Hoping to reduce cultural conflict (and perhaps in self-defense), the French colonists decided to teach the Totonacas the new technique. And before long Mayan vanilla harvests indeed increased as a result of this innovation.

Yet this story should not be taken to mean that the prehistoric and historic Mayans were oblivious to the benefits of crop pollination—they simply encouraged animals to provide this service rather than do the work themselves. The Mayan tending of stingless bees—and the symbolism associated with this tradition of animal husbandry—is a remarkable testament to their ecological insights into the often hidden or forgotten relationships between crops and pollinators. The Mayans of the Yucatan peninsula and adjacent states have long hunted honey from at least 17 different species of native stingless bees. More to the point, four species of native bees have also been kept in their dooryard gardens as semi-domesticates. These four stingless bees, known collectively as *xunan kab* in the Mayan language, pollinate over 200 species of plants around indigenous villages in the Mexican states of Yucatan, Campeche, Quintana Roo, Tabasco, and Vera Cruz. They are effective pollinators for no fewer than 16 agricultural crops grown in the area, too, including coffee, cardamon, avocado, lime, mango, nance, and several additional tropical fruits and nuts.

Gardeners no doubt achieve higher fruit yields when they keep hundreds of stingless bees as managed colonies. Most managed colonies of these bees are kept inside 10-inch-diameter hollow tree trunks in Mayan dooryard gardens. These gardens are rich genetic reserves of folk crop varieties of at least three dozen fruits and tubers, four dozen herbs, six fiber and utensil plants, ten trees yielding stakes and poles, and twenty ornamentals.

The Mayan tradition of caring for native bees was well established in prehistoric times, as the sacred Mayan texts known as the *Chilam Bilam* attest. The fate of humans and bees are often intertwined in Mayan stories and rituals. When a beekeeper dies, for example, the inheritor of his hives must immediately go to the bees and tell them of the death, assuring them that they will remain cared for. The new beekeeper must not visit the cemetery, or help prepare the corpse for burial, for fear that he will bring sadness back to the hive. Any incidental contact with death must be followed by a ritual washing of the beekeeper's hands and arms before he dares to touch the hives. And if a bee is accidentally killed, it is tenderly folded into a leaf, then buried. Of all the natural resources stewarded by the Mayans, only maize is regarded with more elaborate ritualistic outpourings of affection and respect than are the stingless bees.

Even though Columbus marveled at the wax derived from these stingless bees, other European colonists felt that Native Americans were missing something vital by lacking domestic honeybees. By the 1620s, honeybees had been intentionally imported to several places in the Western Hemisphere numerous times. Before long the introduced

*Today, in the Mexican states of Yucatan and Quintana Roo, the contemporary Mayan Indians practice meliponiculture — the keeping of stingless bees (Melipona beechei, Trigona spp.) — as they have done for hundreds of years. Here a beekeeper tends to his Melipona bees in hollow logs by hanging a dish of honey as an offering to them.*

bees began to compete ecologically with the native *Melipona* and *Trigona* bees, but they did not become a major economic resource in the Yucatan peninsula until the twentieth century. Despite this competition from apiculture, meliponiculture, the cultivation of stingless bees, has persisted among the Mayans, including the vanilla-growing Totonacas of the Vera Cruz region.

There are a number of cultural reasons why the Mayans did not abandon their native bees, even though domestic honeybees produce more honey. For one, the native honey was considered to be more effective in treating cataracts, conjunctivitis, chills, fevers, heartburn, laryngitis, and complications at childbirth. Mixtures of pollen and honey from stingless bees have a distinctively sweet yet acidic flavor favored over that of *Apis*-produced honey. The native honey is routinely added to *atole* porridges and squash seed condiments; introduced *Apis* honey is not. The native honey neither ferments nor is crystallized under rustic storage conditions, so it may keep its flavor in storage for two to three years without special precautions.

Most important, perhaps, were the ceremonial uses of honey from stingless bees, ceremonies for which honey from European bees simply would not do. Nevin and Elizabeth Weaver were perhaps the last out-siders to witness a Mayan rite to regenerate stingless bees, a ceremony called the *Hanli Kol*. Formerly enacted every four years under the alter-nating sponsorship of various beekeepers in a village, its celebration has become very infrequent in recent decades. To reinitiate this custom, the Weavers had to sponsor a shaman from another village to come and perform the 24-hour-long rite. Once he arrived, however, hundreds of local Yucatecan Mayan residents were eager to participate. To be per-formed properly, the *Hanli Kol* exacted more than a hundred hours of preparatory work and required the donation of dozens of products from native plants and animals.

This ceremony, intended to allow a local beekeeping family to double the number of colonies it kept, alternated between ritual prepa-ration, the sharing of native foods, and the chanting of an ancient prayer seven times. The prayers in the rite had been directed toward *Noyum Kab,* the Great God of Bees, who came from the traditional home of native bees, *Yal Koba,* a classic Mayan temple. Ironically, when European honeybees were first introduced to the village decades before, they caused great excitement, for many of the local Mayans mistook them for the envoys of the Great God of Bees mentioned in the cere-mony. Gradually, however, they realized that honey from European bees had none of the curative properties ascribed to native honey. From then on, keeping European honeybee hives was considered an eco-nomic pursuit devoid of any redeeming cultural value. Traditional Mayan ceremonies have no power over the introduced apiaries of Euro-pean honeybees, or over the omnipresent Africanized race now domi-nating the area.

Historically, there was indirect suppression of Mayan beekeeping because of the Spanish distrust of the "witchcraft" that went along with it. Mexican-American poet Francisco Alarcon reminds us of this: "'Bees are godly servants of the flowers/they keep for themselves/they make the wax we burn to our Lord/for that we love them we revere them'/said Miguel the bee seeker after being tricked to recite/the incantations of the beehives he knew better than his Ave Marias."

Unfortunately, the number of stingless bee colonies tended by the Mayans has declined as more people pursue the purely economic rewards of European honeybees, investing less time in the native bees. The number of tended colonies of stingless bees is now half of what it was in 1980. Today, there remain only 530 families tending stingless bees on the Yucatan peninsula, and these Mayans collectively maintain fewer than 5,000 colonies, largely of the most easily "tamed" species, *Melipona beecheii*. Deforestation has opened up sugarcane and henequen plantations in surrounding areas, plantations which use insecticides that have virtually eliminated stingless bees from certain stretches of the coastal plains. The gene pools of these bees are in steep decline, and so is the traditional pool of cultural knowledge about them.

In the midst of deforested areas, some Mayan farmers who have yet to abandon this ancient beekeeping tradition are realizing that their honey yields and crop yields are now unusually low. Some blame the decline in honey on a scarcity of floral resources in the logged-over forests adjacent to their home gardens. Others blame both crop and honey declines on local droughts, which follow in the wake of deforestation. Curiously, certain Mayan farmers blame this lack of rain on their own failure to continue the rain-bringing ceremonies their forefathers had always offered—ceremonies which require that honey from the *xunan kab,* the stingless bee, be left in their fields. Without that ritual observance, the harvests will suffer.

Mayan gardeners must therefore maintain the native stingless bees—honeybees are not enough—in order to sustain the rituals that ensure good harvests. Metaphorically, at least, the Mayans have recognized the critical connection between native pollinators and fruitful harvests. When deforestation wipes out nectar plants, and the few select trees that offer trunks suitable for beekeeping, the feedback loop unravels. It is frayed, too, by cultural changes that keep younger Mayans from becoming gardeners and farmers who maintain the ancient thanksgiving rites associated with stingless bee honey.

Today, as never before, Mayan meliponiculture is at a crossroads. With the arrival of Africanized honeybees in the Yucatan, the peninsula's sizable industry based on European honeybees is in peril. In recent decades, Yucatan had become the largest honey producer in Mexico. Today, however, crops that depend on European honeybees for pollination are in a precarious position. If native stingless bees ever had an opportunity for revival, it is now. Yet few Mayan youth see the tradition as attractive, even if innovative technologies make it less labor-intensive. In their eyes, it is too reminiscent of the "old ways" from which they are attempting to escape.

Even against such stiff odds, the Mayan beekeeping traditions are entering an era of revitalization. Through the urging of a remarkable young ethnobiologist, Sergio Medellin Morales, the MacArthur Foundation has recently funded the formation of a new nonprofit organization, the Asociación Civil *Yik'el Kab,* named for "native insects that produce honey." In February 1991, Medellin convened this dynamic coalition of Mayan beekeepers and scientists to draft a strategic plan with several objectives: to conserve and expand stingless bee production, to open new markets among health food stores and herbal medicine outlets, to sponsor the planting of more dooryard gardens and rainforest restoration plots filled with nectar plants, and to protect old-growth trees suitable for feral bees and the elaboration of colony-bearing containers.

This project has had the technical assistance of some of Mexico's finest experts on sustainable development, including Arturo Gomez-Pompa, Enrique Campos-Lopez, and Jorge Gonzalez-Acereto. More important, perhaps, at least 20 Mayan beekeepers have served as consultants at training and planning workshops, while hundreds of others have been surveyed to voice the problems and their solutions. If any ancient ecological practice has a chance of persisting in the face of current challenges, Mayan meliponiculture may be it.

But, as the Weavers realized while living among their Mayan neighbors, "the actual work with [native stingless] bees cannot be separated from the rituals which accompany it." It is not merely a novel biotechnology; it is a living culture. Should anyone fail to honor the rituals to which the bees have become accustomed, the bees may lapse into an unproductive state. That is why talking with the bees in their dooryard gardens is the Mayan beekeepers' most time-intensive activity: staying in touch, keeping an ancient flame—the dialogue between different species—well kindled and burning into the future.

S TEVE REMEMBERS:
I was on the road again, this time headed toward the "sky island," the Chiricahua Mountains near the New Mexico–Arizona border. I passed mile upon mile and row upon row of apple trees—Red and Golden Delicious, Jonathans, Fujis, and Granny Smiths—before breaking out into a plowed but unplanted area of almost 200 acres. There in the middle of it stood an immense greenhouse that sheltered one vision of what managed pollination will be for many crops in the future.

I had never seen such a glass-enclosed structure before. In amazement I scanned the length of a single greenhouse encompassing 10 acres—three times the size of southern Arizona's famous New Age tourist attraction known as Biosphere II. The glass oddly distorted the sky around and above it in the Elfrida Valley, one of Arizona's last bastions of "dry farm" homesteads earlier in the century. Where desert-adapted native tepary beans once grew, the land and the fossil groundwater below it were now considered too expensive to justify rain-fed staple crops. Inside the glasshouse structure was a world of hybrid tomatoes and fuzzy black and yellow bumblebees.

After I had parked, I was met at the door of the corporate offices by a security guard, who immediately phoned the resident manager of Bonita Farms. The manager remembered me, for I had arranged my visit months in advance. No competing agribusinesses, no chambers of commerce, no Future Farmers of America were allowed access to this high-tech, computer-run tomato factory. Security checks abounded. I was grateful, though, that they let me visit the impressive facility. The nearest technology of this type exists only in the Netherlands and a few other European countries. So, in a way, it was like stepping inside a Dutch greenhouse.

And then the tour began. The factory, they said, efficiently captures solar energy but must nevertheless use fossil fuels to control the microenvironment through cooling and heating. It felt as though we were strolling through the lungs and circulatory system of an imposing robotron. Tens of millions of dollars of venture capital had already been invested in developing this operation. But that was peanuts compared to what it would take to run the system when taken to its logical end point. The corporation was planning to put another 10 acres of land under glasshouses each year, until the full 200 acres of plowed ground surrounding the present structure had become one vast controlled microenvironment.

"Where's the soil?" I mumbled, a bit confused. The manager pointed to the white plastic bricks from which each tomato plant emerged. None of them were filled with the good earth of the Elfrida Valley. Instead, they contained rock wool, sensors, and emitter tubes. Upon demand—and monitored by the water-stress sensors—they were automatically watered and fertilized.

I looked down at his feet at some thin white pipes, and again scratched my head. These conduits, he told me, transported hot water to keep the entire greenhouse at optimal temperatures for rapid plant growth and tomato maturation. The same white pipes doubled as runners for electric trolleys that carried "farmworkers" between the rows, where they pruned, tied, and harvested plants at different stages of growth. The local Hispanic residents who dominated the meager workforce found the tomato factory far more lucrative, and far more regimented, than the "dirtier" fieldwork of their youth.

There were actually few tasks that needed to be done by humans in this place. Above their heads were, not clear skies, but electric cables, blinking lights, and display panels attached to electronic data-loading stations. Those stations fed temperature, humidity, and nutrient data back to centralized computers that monitored overall conditions and timed the delivery of controlled substances to each plant.

I looked out at the cutting edge of European glasshouse engineering transplanted to American soil, hardly knowing what to say. The manager was quite proud of the weed-free environment, for it meant that his tomatoes could be grown in a herbicide-free environment. In fact, the lack of weeds also kept down the presence of the few insect pests that happened to arrive in the glasshouse, so that no chemical insecticides were required either—biological control agents kept these few stragglers in check.

This was all well and good, since the operation relied on pesticide-sensitive bumblebees for pollination in order to produce large tomatoes free of any blemishes. The manager boasted that, in fact, the tomatoes "were organically grown." I had to wonder whether rock wool, plastic drip emitters, and microsensors could qualify as organic materials.

At last I asked to see the bumblebees, the real reason I wanted to visit Bonita Farms. I was in luck—a shipment of *Bombus occidentalis* colonies had just arrived by air cargo to Tucson. It had been driven to the farm already, so I would have ample chance to see them in action. These bumblebees were purchased on a monthly service plan from an outfit in central California with ties to a multinational based in the Netherlands. The colonies were rented on the basis of their capacity to pollinate a given "volume" of greenhouse tomatoes. The bees had been raised in total isolation from truly wild bumblebees, under rigorous hygienic standards, using the latest high-tech insectary methods.

I found each bumblebee colony sitting on a platform a full 6 feet above the broad central concrete walkway: a living shrine amidst all the wire, glass, and plastic. The boxes were emblazoned with the black and yellow "Natupol" trademark and the imprimatur of the Dutch-owned Koppert Biological Systems, a multinational corporation largely responsible for the technological advances in bumblebee breeding and rearing during the past decade or so of frenetic growth of this largely European industry.

During my first visit to the Bonita glasshouses, I had been advised not to venture too close to the bee domiciles, let alone peek inside. Over the intervening years, however, security had somewhat relaxed so that the "secrets" associated with the gaudy black, yellow, and white boxes were now more widely known. *Bombus* colonies had recently been sold to individuals unaffiliated with Natupol's corporate family. The bee was out of the bag, so to speak.

You could easily unplug the round exit hole of the nesting box to gain entry into the world of managed bumblebees. The colorful outer packaging simply hid some cleverly molded styrofoam walls, a layer of inner plastic covers, drainage ways, and a plastic gravity-fed sugar water feeder borrowed from the poultry industry. Atop the hive was a clear plastic bag, nearly identical to the ones used in hospitals to hold human plasma for intravenous feeding. Here it held not blood but an amber-colored solution of artificial nectar known to Bonita Farms and other users only as "Bee Happy." Whatever was in this secret recipe formulated by Dutch masters, it worked. The bumblebees certainly seemed to enjoy slurping up this nectarlike elixir in the safety of their colonies. In this nectar-free greenhouse environment, providing the bumblers with "Bee Happy" juice was absolutely essential if they were to pollinate the thousands of hybrid tomatoes around them.

Like many plants that require *sonication* by bees—plants including blueberries, cranberries, eggplants, and kiwi fruits—tomatoes produce nutritious pollen but not a drop of nectar. Blossoms of tomatoes and many of their nightshade relatives can produce some self-pollinated fruit, but they are much more prolific and produce well-formed, higher-quality fruits when they are sonicated, or *buzz-pollinated,* by certain bees that perform a peculiar "rotating dance" on these yellow and green starlike flowers. But the bees that gain pollen by buzzing flowers must either have other flowers available as nectar sources or they must be mainlined a nectar substitute. High-tech bee ranchers are more than willing to treat their lilliputian livestock to "Bee Happy" in exchange for the services they offer.

Early in the morning, within their five yellow petals, the open flowers reveal a turgid central cone of anthers ripe with cream-colored pollen. Any bumblebee that chances upon this payload knows exactly what to do to get it. Although tomato flowers have no appreciable scent, the contrasting colors of the newly opened blossoms guide the bees in close enough to determine whether a pollen load is still present.

If you focus on one plant's flowers, you will begin to hear the low-pitched drone of wingbeats and then see a large worker bee alight on a cluster of virgin tomato blossoms. Choosing one, she will grab it tightly with her mandibles and legs, then curl her abdomen into a tight "C" shape around the bright yellow cone of anthers. Suddenly her body will blur in rapid motion. There will be an immediate undulation of her midsection—her thorax muscles are capable of powerful contractions—and as her body shivers, a soft pulsing buzz can be heard. You will detect sounds approaching a middle C—512 cycles per second—as you watch her shudder while perched atop the tomato blossom.

These vibrations are good enough to cause pollen grains to dance dizzily within each hollow anther in the cone. At first, they will only stir inside the anther walls, occasionally bumping into each other. Then, as the bee vibrations intensify, the pollen becomes so agitated that it begins to boil and stream out of slits in the anther tips like a liquid slurry. As the worker bee gyrates ever more powerfully, the pollen will fly in a dense cloud before her. Thousands of pollen grains will spill out and stick to the undersides of her hairy thorax and abdomen. After a few seconds, the bumblebee suspends her buzzing, loosens her grip, reattaches to the flower, and then begins a second bout of buzzing. The worker will repeat this behavior over and over just a few inches from your face. Intent on amassing a large pollen load to help feed the colony, she will perform hundreds of such buzzing bouts until her corbicular "bee baskets" are bulging with pollen.

Now you begin to watch another bee worker pausing between foraging forays. Hanging by the tarsal claws of one leg, she is making rapid grooming movements with her other legs—quite a balancing act. Her entire body has been dusted with a light rain of pollen, so she has decided to rake and brush it off into the pollen baskets on her hind legs for easy transport back to the nest. Since many of the bees are taking these grooming breaks, the tomato flowers must be bursting with pollen this morning.

S TEVE REMEMBERS:
Later in the morning, just before I was to leave Bonita Farms, I took one last look around the greenhouse. So close to noon, there were no longer any bumblebees foraging among the flowers. I glanced around, spotted what looked like a fresh, still-turgid yellow flower, then pulled a tuning fork from my pocket. I instantly twanged it against a metal support, and then touched it to the cone of the tomato blossom. Almost no pollen was discharged. If this had been a virgin flower, an explosion of pollen would have formed a cloud. I twanged the tuning fork again, and tried another promising flower. No cloud. I could only shrug. I conceded that the bumblebees had efficiently harvested the morning's standing crop of pollen.

If you had visited a glasshouse such as this a decade or so ago, you would not have encountered any bumblebee workers—only humans with battery-powered vibrators in hand. Long before European glasshouse tomato growers had pioneered the art of bumblebee keeping, they were well aware that superior fruits and higher tomato yields could be achieved when each flower was physically touched and shaken. Workers would carry giant vibrators, especially constructed of wood and metal, up and down the aisles between glasshouse benches, mechanically achieving much the same results that bumblebees had achieved on nightshade flowers for tens of thousands of years.

It was just too labor-intensive, however, and prohibitively expensive, given the other costs a tomato factory owner inevitably faced: infrastructure, heating, cooling, fertilizers, and, in many cases, frequent pesticide applications. Growers tried vainly to find other means of replacing bees with quick, cheap, mechanized fixes: shaker tables as benches for tomato plants, for instance, or huge speakers blasting out high-energy sound waves to acoustically force pollen from the tomato's anthers. At last, they admitted that none of these mechanical methods were as efficient as the touch and shudder of a live bumblebee. By the mid-1980s, the growers finally began to have successes in keeping bumblebees in glasshouses, thanks to refinements in the quality of artificial nectar and innovations in the design of domiciles. A multimillion-dollar industry of brooding bumblebees caught hold in western Europe, and within five years had spread to other continents.

Yes, the industry had invented a new means of beekeeping, as Eurasians or North Africans had done with honeybees in the Old World, and Mayans or Aztecs had done with stingless bees in the New World. Yet each of those older forms of beekeeping was rooted in a respect for the magic of bees, and imbued with rituals to keep their human stewards humble and grateful that their relationships worked. In short, both ancient apiculture and living Mayan meliponiculture had cultural manifestations that guided them pragmatically, ethically, and spiritually. Bombiculture, however, had no such cultural context in which to operate. Bumblebees had replaced battery-powered vibrators—surrogate mechanical bees—but were nevertheless treated as just another high technology, albeit a biotechnology. For most glasshouse tomato growers, *Bombus occidentalis,* along with other species being "domesticated" at the present time, was simply a more cost-efficient keeper of the vibrations than any human worker had been.

S TEVE REMEMBERS:
I stopped at the last tomato plant at the end of the row before the door, to take a look at just one more tomato flower. I took out my tuning fork and tried my parlor trick again. The twang did not entice any more pollen from the anthers. Checking the flower with my hand lens, I noticed some telltale brown markings on the otherwise bright yellow anthers.

"What are these brown spots?" I asked the manager.

"Oh, those? We call them 'bee kisses,'" he replied. "Those are the places where bumblebees bite into the anthers to keep from getting thrown off the flower while buzzing their brains out. You know, it's like the browning you find when you bite into an apple, and then you set it aside for awhile. That's how tomato growers can be sure that the bumblebees have been doing their work."

As I left the greenhouse that afternoon, I tried to imagine what tomato growers might be like in two decades—or even twenty decades from now—after having spent season after season diligently searching their plants for bee kisses. Would the sweetness of that metaphor get under their skin, skin which is sometimes as tough as that of the hybrid tomatoes they grow? Would songs develop in praise of bee kisses or their furry black and yellow go-betweens, rituals expressing gratitude for the forthcoming harvest?

I reserve judgment. Nonetheless, it may become less and less possible for tomato growers to treat the bees without some affection, some respect, or some sense that they are members of a new "agrisymbiosis," not merely links in the mechanistic chain of some ultramodern food factory. Working in a place where they are literally surrounded by bee-kissed blossoms, how could they remain unmoved?

*A modern U.S. beekeeper, wearing a mesh veil and hat to protect his face from stinging guard bees, inspects his honeybees* (Apis mellifera) *in a Langstroth hive, technology invented at the time of the Civil War. He uses smoke to calm the bees but often wears no protective gloves as a way to avoid crushing bees and releasing alarm scents.*

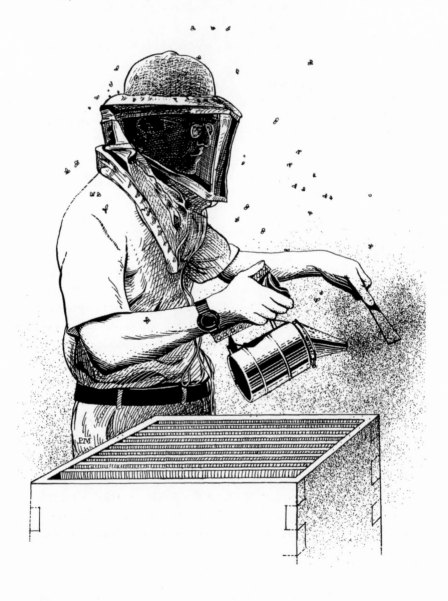

# CHAPTER 10

# New Bee on the Block

## Competition Between Honeybees and Native Pollinators

G ARY REMEMBERS:

In the fall of 1993, I walked with my daughter around Colonial Williamsburg, Virginia, amazed at the ingenious ways the National Park Service had found to remind us what early colonial life was like. The Park Service staff was all dressed up in period clothing, vigorously acting out their colonial tasks. They had chosen to portray the various crafts and cottage industries they associated with Early (Anglo-) American life on the Eastern Seaboard. As the November days proceeded toward Thanksgiving, they hauled in the crops, hung the herbs to dry, made a new batch of candles from the tallow of recently butchered stock, and stacked firewood to be used in the foundries that winter. The women hand spun and wove wool blankets in colors that matched those of the fall leaves: rust, burnt orange, crimson, and ruddy brown. They were all working with the single-mindedness that many colonists express as they settle in a new land and come to dominate it. And yet, I kept asking myself, "What did their presence mean to Native Americans? How did they change this world?"

It had been 375 years since the introduction of honeybees into North America at Jamestown, Virginia, and their establishment in other colonies like Williamsburg nearby. Over that timespan, Native Americans greeted Europeans, then succumbed to their diseases. Jamestown and Williamsburg fell into decay, then became National Historical Sites. Red wolves were driven into extinction on the Eastern Seaboard, but have recently been reintroduced to Virginia. Over those same 375 years, however, honeybees have spread into every conceivable terrestrial environment on the North American continent. They now number in the billions.

To my mind, the honeybee is a far more classic example of a colonizer than those at Colonial Williamsburg, although both fit the roles of an exotic invader on a trajectory of ecological conquest. Few colonizations have been so successful as those of the honeybee. Since Jamestown, purposeful introductions of many other races of *Apis mellifera* have occurred in all corners of the earth. The acceleration of migratory beekeeping along highways built since the 1940s have moved honeybee colonies into every exploitable niche in the continental United States. They are more ubiquitous than the cow, the horsefly, the tumbleweed, and the tax collector. And yet few of us have asked, "What does their presence mean to the natives? How have they changed the world?"

It may be too late to gain a sense of what America's pollinator landscapes were like before the honeybee's arrival. All we can do is wonder how many beasts it has eliminated in its wake. Many of the most serious impacts of honeybee colonists on native bees and plants probably occurred during the first 50 years following their introduction and spread westward. And that is the trouble with assessing their impact by hindsight. Today, it is virtually impossible to find any area of the world free of feral or managed bees where we might learn what that first collision between natives and invaders was like. This makes it tough to demonstrate that other bees have been the victims of *scramble competition* by honeybees—the disruption of many plant/pollinator relationships by a single dominant exotic. As community ecologists have been quick to point out, we have no "control" or well-documented benchmark for comparison. There are few "pristine" environments free of honeybees.

Nevertheless, studying the impact of honeybees—including Africanized races—on indigenous New World species has been the consuming passion of several scholars. These ecological historians and field scientists suggest that *Apis* is often able to outcompete native species such as bees, ants, and wasps. The honeybee does so by virtue of its superior ability to detect, to direct its peers toward rewards (via dance language), and to rapidly harvest floral resources. It can scout out even the most ephemeral patches of pollen and nectar. It can also quickly decide whether an individual flowering tree is worth mounting an expedition over. But do adept honeybees truly usurp resources essential to the survival of native bees? The notion of scramble competition has been the subject of several illuminating studies.

Two decades ago, University of Arizona ecologist William Schaffer began to evaluate competition for nectar between introduced honeybees and several natives, bees and ants included. He focused on the pollination ecology of the native shindagger, *Agave schottii,* a century plant with knife-like leaves. Shindaggers have a short flowering stalk bearing dozens of fragrant, tubular yellow flowers. They are pollinated by a variety of native bees including carpenter bees, bumblebees, and smaller kinds. Although originally interested in how the agave's "big bang" reproduction affected native pollinators, Schaffer gave more and more attention to honeybees at his Santa Catalina mountainside sites.

In 1979, Steve joined Schaffer for these studies, introducing genetically marked ("blonde") Cordovan honeybees and pollen traps to the observations already under way. Nectar volumes—measured as standing crop in open flowers and total harvests in bagged flowers—were assessed along strips of plants that were "tanglefooted," to exclude foraging ants, or left untreated. Steve also introduced colonies of honeybees after observing patterns of nectar use by native bees and ants for several weeks. Bill and Steve continued to take pollinator censuses at the flowers, as well as nectar measurements, during the hives' entire residence time and even after their removal.

A few feral honeybees were already present on the site, but the researchers introduced colonies of genetically distinct Cordovan honeybees to determine the effects of large numbers of honeybees on the native bees. What did they find? That honeybees predominated at the shindagger sites which had the most nectar per plant. And that honeybees preferred to forage at those times of the day—early morning and late afternoon—when floral resources were most abundant. In short, the introduced honeybees usually dominated the pollinator fauna on the most productive sites, but natives were more resourceful elsewhere. This finding is undoubtedly related to the honeybee's exorbitant energy demand for floral sugars. Honeybees have become masters at finding and then communicating the distance and direction to the most profitable floral patches. Once these patches are found, they rapidly and efficiently swamp out the other bees—not through overt aggression, but by sheer numbers. The natives hold their own only in less productive patches.

The story gets more complex from there. When the two Cordovan honeybee colonies were trucked in to the study site, they had no effect on the numbers of feral honeybees on the stalks. This indicated that the local population of feral honeybees was large enough to resist being overwhelmed. At first the new bees on the block were allowed to forage freely to see if they would reach some equilibrium with their competitors within a few days. Then the two Cordovan honeybee colonies were taken away.

That was when Schaffer's team noticed that the number of feral honeybees observed on flowers increased until it approached the previous combined totals of light-colored Cordovans and the darker feral honeybees. The team also documented that the floral visitation levels for the bumblebees and small halictids and other bees increased at first, then declined. Although these experiments conducted by the Schaffer team were exacting in detail and complex, some crucial points emerge from their multiyear studies. Above all, even against a background of dark feral European honeybee colonies among the rocks, the introduction and subsequent removal of genetically recognizable Cordovan stock had a strong effect, not only on the native bees and ants, but also on foragers from the wild honeybee colonies living nearby. In Arizona at least, Schaffer and others have not found any honeybee-free study sites to completely eliminate the influence of *Apis* foraging.

They learned that moving even a few large honeybee colonies into a natural habitat can have immediate and far-reaching consequences on the guilds of nectar and pollen-feeding animals present—and on the reproductive and fruiting success of native plants such as the shindagger *Agave* they examined. Their ambitious experimental regimen was cleverly designed to document shifts among nectar-feeding bees, ants, and wasps on their tagged flower stalks over a few days. These and related studies have given us undeniable indications that as exotic interlopers, honeybees do have an immense impact on local (native) nectar sippers.

Bernd Heinrich—a noted pioneer in insect foraging energetics and a popular natural history writer—has produced calculations to explain honeybee competition with native bumblebees. Working for decades with native bumblebee colonies living and foraging in East Coast sphagnum bogs, Heinrich realized that bumblebees lead a highly precarious hand-to-mouth existence. Unlike managed and feral honeybee colonies, which hoard massive amounts of fossil fuel as capped honey stored in their waxen combs, smaller bumblebee colonies are barely able to sip from the blossoms fast enough to feed themselves. They are typically left with a tiny amount of precious surplus to feed their brood. If they are kept at home on too many rainy days, the colony will starve. Remember hearing about oceangoing honeybee colonies locked inside casks on the ship? They were living off their own hoarded "fossil" nectar deposits. Bumblebees have no such honey reserves to fall back on.

Heinrich reasoned that the effects of scramble competition for limited floral resources between introduced honeybees and native bees could be deduced from a rough calculation. A California colony of the bumblebee *Bombus vosnesenskii,* he found, utilized about 1.9 ounces of honey and 1.2 ounces of pollen every day. From this measurement of food intake and from energy investments of workers and queens, Heinrich estimated that it takes about 5.3 pounds of honey and 2.6 pounds of pollen for a bee to reach the mature nest stage. At that point, it would have produced 375 reproductive progeny (males and virgin queens).

Heinrich considered this to be a *minimum* calculation of the resources required for the colony to continue. He concluded: "For every gram of honey produced by honey bees in an area suitable for bumblebees, there would be about 0.16 fewer bumblebee reproductives produced. A strong hive of honeybees in the United States collects on the average about 441 pounds of honey for its own use, and about 88.2 pounds surplus that can be collected by the beekeeper. A single honey bee hive, then, could hypothetically reduce the population of bumblebee reproductives by 38,400." That is the equivalent of 102 bumblebee colonies. Heinrich saw the writing on the wall, and for many bumblebee populations it was an epitaph:

Men have provided food for them by clearing the land, cultivating crops, and introducing many flowering weeds. However, when honey bees are present in areas with native vegetation, where wild bees normally harvest nearly all of the available nectar and pollen, they adversely affect the populations of bumblebees and other wild pollinators. Decreases in wild bee populations appear to be directly proportional to the amount of nectar and pollen made unavailable to them, although at the present time there are no data to confirm or refute this hypothesis.

Taking Heinrich's message to heart, we tried to see if it could be confirmed in the Sonoran Desert region of Arizona. There we discovered that honeybees could collect vast amounts of pollen from any one of several flowering plants that a carpenter bee might depend on for its entire livelihood. Managed honeybee colonies operated by beekeepers at an average site often harvested a hundred pounds of pollen each year from the floral riches in ancient desert legume forests. In contrast, a female carpenter bee tries to wrestle 0.1 pound or less of pollen from the desert in order to stuff her nest chambers with some 20 to 30 million pollen grains each for her progeny. Each honeybee colony nearby is eating her progeny out of a life by consuming the "nest equivalents" of perhaps 200 to 400 carpenter bee nests scattered over many square miles of rugged desert hillsides. Clearly, honeybees are in direct competition for food with natives like carpenter bees and bumblebees, for there are only finite daily supplies of pollen and nectar available from native desert plants. Honeybees are literally taking the food right out of the mouths of babes.

Community ecologists will probably argue for many more decades about how much competition truly exists in nature. One thing they generally agree on, though, is that if two species compete for a food resource, such as nectar or pollen from flowers, then this resource becomes in limited supply. Further, one species may be better at harvesting resources than the other—presumably diminishing the other's harvest. Thus, the evidence for direct competition between two or more species in nature is usually circumstantial. Very few field studies have ever

*Common types of pollen grains from flowers pollinated by bees include those that are highly ornamented with spiny surfaces such as dandelion* (Taraxacum) *and chrysothamnus.*

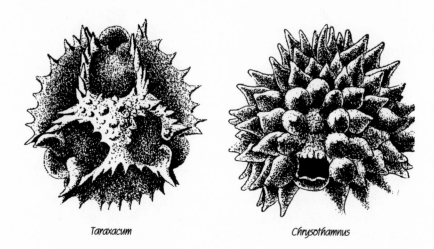

Taraxacum                                    Chrysothamnus

demonstrated incontrovertible proof for competing animals. The Old World honeybee *(Apis mellifera)* is generally thought by ecologists and entomologists to displace native pollinators from both floral resources and geographic areas. It does so not by aggressive interactions at flowers—though some species of stingless bees do fight over flowers—but by overwhelming all the other bees with the sheer force of their populous colonies and rapid sequestering of ephemeral food sources.

Dave Roubik of the Smithsonian Tropical Research Institute in Panama has been carefully watching for honeybee competition with native bees for more than 17 years at the very same sites in Latin America where he began with them. Over that period of time, many of the virgin stands of tropical forest have been cut down around him, so that secondary forests, roadsides, and fields have become more common in the landscape. He has documented how pollinators respond to these landscape-level changes, and his results are sobering. As early as 1978, Roubik was publishing data on introduced honeybee and stingless bee abundances on flowering plants of *Melochia, Mimosa,* and *Rhynchospora* in lowland tropical habitats and inferring competitive advantages for the newcomers. In long-term experiments conducted in French Guiana and Panama, Roubik manipulated numbers of honeybee colonies in areas where there were few or no bees at the start of the studies. He also conducted a multiyear study of honeybee and native bee visitors to flowering patches of a sensitive plant, the pink-flowered *Mimosa pudica.* He predicted that shifts in resource use caused by colonizing Africanized bees may lead to population declines in some groups of bees and other Neotropical pollinators.

Now, almost 20 years later, and with many additional studies, Roubik says that "the honey bee's spread and apparently permanent colonization of tropical America may be likened to a vast experiment," since they now occupy approximately 20 million square kilometers of territory from northern Argentina to the southwestern United States. And the Africanized honeybees aren't likely to leave. In fact, they are likely to have a powerful impact on other bee species, along with other pollinators and their food plants. Although Roubik found essentially no adverse effects on the long-term stability of certain other apids—such as the metallic honeybee orchid bees—he did demonstrate domination at certain flowering patches over social stingless bees. He also documented intense but sporadic competition for food by Africanized honeybees that actively burst forth for short periods, a few hours or days, enabling them to exploit floral resources much faster than other bees on the same flowers. Furthermore, he made a dire prediction: Africanized bees might cause the local extinction of several species of *Melipona* and *Trigona* stingless bees after periods of coexistence and resource overlap lasting more than ten years.

Wherever roadbuilders blaze a route into untouched tropical forests, woodcutters and Africanized bees are among the first to travel those roads. The introduced honeybees rapidly colonize first the roadsides, then the new forest clearings. Later, they invade the secondary growth that comes up in abandoned slash-and-burn fields. Finally, they begin to invade the remnants of rainforest stands, often competing with, but seldom displacing, the natives remaining there—natives that have since abandoned (or been sprayed out of) the disrupted habitats. Roubik's long-term studies in Central America confirm what biologists on other continents have long suspected: honeybees are a disturbance-loving species that first follow human disturbances and then create some of their own. But is this true only for bee-to-bee competition, or for honeybee competition with other pollinators as well? We turned to Australia for an answer.

A distant 65 million years ago, set adrift from its tectonic plate traveling companions, the young Australian continent carried with it a truly unique bestiary. These stowaways included atypical sets of bees dominated by colletids and mammals dominated by marsupials. Many of them turned out to be avid pollinators. Today, after so much isolation time protecting them from competition with other nectar-feeders, Australia's native pollinators face stiff challenges from newcomers. Nevertheless, the continent boasts the largest and most diverse assemblage of colletid bees of any continent, many found nowhere else. Although we emphasize the adaptive radiation of many colletid bee groups (especially *Leioproctus*), there are many representatives of all the other major bee families on this continent. Other bee families are notably poor or altogether lacking "Down Under." This does not mean that Australia is depauperate in native insects and vertebrate pollinators. Rather, its age and its long separation have produced a unique assemblage of floral visitors that includes at least ten families of flower-visiting beetles and five families of nectar-feeding birds, along with many syrphid and muscid flies and bees.

Once, in Western Australia, we heard a lecture that set our minds spinning into another realm of the competitive effects. We were attending a seminar in Perth given by Professor R. Wills on the subject of the nectar and pollen plants in the extremely diverse sand heath plant community known as the *kwongan*. During this lecture, we learned that honeybees were taking large amounts of pollen and nectar away from native wildlife. Two hundred years ago, the Australian continent was irrevocably disturbed by the arrival of English colonists bringing with them a few souvenirs of their British homeland: rats, sheep, cattle, and various birds. One of the animals brought Down Under was the European honeybee, one more means of domesticating an unfamiliar landscape, reshaping it in the image of Northern Europe. In fact they had gotten into direct competition with endemic nectar feeding birds as well as native bees.

Until quite recently, Australian honeybees have been appreciated by all segments of the human population, not just beekeepers. Farmers and conservationists alike were in love with the honeybees since they are commonly assumed to provide essential pollination services for both cultivated crop plants and native plants. This view has been undergoing a radical overhaul. As more studies are performed on scramble competition between the introduced bees and their endemic counterparts, the honeybee is looking more like the guy wearing the black hat. Today, ecological researchers and conservation biologists have good evidence that honeybees are adversely affecting vertebrate pollinators including rare honeyeaters and honey possums. Especially during critical breeding periods and during years of drought, nectar and pollen production in the bush are so low that native nectar-feeding wildlife starves while honeybees usurp their needed floral resources.

From an ecological and evolutionary perspective, honeybees may pose a hitherto unsuspected threat to life in the bush. Not only do they displace native pollinators—both insects and vertebrates—from flowers but they do so without effectively triggering the pollination mechanisms of the crop plants or native flowers they visit. A significant portion of the Australian flora requires vibratory pollen harvesting to set fruit—buzz pollination—and honeybees are incapable of this feat. Blooms of deadly nightshade blossoms and native bush tomatoes all require other pollinators to effect fruit set. Whereas bumblebees and carpenter bees do an excellent job and routinely set large fruits full of seeds, honeybees leave these plants unfertilized. Although honeybees do not use floral sonication to harvest pollen, they are often quite abundant on Australian blossoms with pored anthers and must account for *some* pollination and seed set. Australia is exceedingly rich in such flowers—especially in the elephant apple family, the Dilleniacae, including the genus *Hibbertia*. Other buzzed flowers are found commonly within the asparagalean-lilioid families (Anthericaceae, Phormiaceae). In fact, buzz pollination is so important in Australia that orchids lacking loose pollen grains have gone to great lengths to produce flowers that effectively mimic buzzed flowers, especially in the genera *Caladenia* and *Thelymitra*.

Australia now has about 525,000 managed honeybee colonies and an unknown number of feral colonies. With the continent's area of 2,969,236 square miles, Australia's half million colonies are placed at a density of only 0.18 per square mile—compared to one managed colony per square mile for the United States. Both migratory beekeeping—trucking bees along highways to blooming crops or patches of native heath or forest in flower—and nonmigratory beekeeping, with long-term established apiaries, are practiced. To make matters more problematic, Australia is the only country in which tracts of native bush are set aside as beekeeping reserves. Their purpose is to provide pasturage for honeybees, for there are other wildlife preserves where beekeeping is expressly forbidden.

This policy has created additional controversies that have raged for decades between sheep graziers and beekeepers. They have fought over such things as legally blocking the government-sponsored release of insect biological control agents to eliminate a pernicious introduced weed. An introduced borage from Europe, known as *Echium plantagineum,* is quite invasive over millions of acres in New South Wales and other states. The plant was purposefully introduced by a gardener to brighten the dry landscape, but like other plants introduced into Australia, it ran amok. And although it is poisonous to certain livestock, such as sheep, it happens to produce abundant nectar. Fine honeys are harvested by beekeepers lucky enough to position their apiaries close by. In short, this introduced plant has engendered much public debate and political outcry between the warring factions.

The plant's common names—used by different groups of distinct persuasions—are quite telling. To the grain farmers and orchardists, *Echium* is "Patterson's Curse" since it is an invasive weed in their cultivated fields. But beekeepers and sheep graziers call the same weed "Salvation Jane." The beekeepers always try to locate their apiaries near the huge purple-flowering patches of *Echium* since honeybees eagerly seek out its abundant nectar and the apiarists can make huge crops of high-quality honey from it. The sheep graziers—even though *Echium* is known to have toxic alkaloids, especially for horses and sheep—still favor the plant. During drought years, Salvation Jane is about the only plant to put out succulent new shoots following rain, and these are eagerly devoured by their valuable Merino and other breeds of sheep.

If it is possible to detect even subtle effects of interspecies competition at flowers for pollen and nectar, then Australia is undoubtedly the best place to look. This is because of the southernmost continent's unique geological and biological history. Australia is populated by representatives of three immense plant families—the Myrtaceae (eucalypts), the Proteaceae (proteas), and the more familiar Leguminosae ("wattles" or acacias)—along with an overabundance of the euglossine colletids, the so-called membrane bees. Although nectar and pollen-rich genera like *Eucalyptus* are rich in species, the largest plant genus in Australia is *Acacia,* known locally as "wattles," with from 700 to 1,100 species. Further, Australian plants and their pollinators have evolved for millions of years not only in isolation but largely in the absence of highly competitive social bees such as camp-following honeybees. The social stingless bee genus *Trigona,* however, was a potent force to be reckoned with by other florivores in the tropical rainforests of Queensland and other far northeastern territories. Still, the native Australian bee fauna had never experienced anything like the onslaught of "white man's flies"—the introduced honeybees from Europe—upon their purposeful landfall 200 years ago.

Australia is also unusual in having a large number of specialist birds, such as honeyeaters and wattle birds, that depend on bird-adapted flowers for much of their energy budget. This is especially true for the Mallee heath and the sandplains of Western Australia. It is no coincidence that Western Australia is also the site of many studies focusing on nectar competition between native birds and introduced honeybees. Some of these bird-pollinated flowering plants even secrete nectar at night that can be worked by the night shift or the early birds and bees the following morning.

Many native Australian plants are used as both nectar and pollen sources by populous honeybee colonies. In Western Australia, Robert Wills recorded honeybees visiting the flowers of 125 species out of an available 413 species within the flight range of his study apiaries. In similar habitats in South Australia, honeybees visited more than 180 out of a total of 360 native plants examined. Honeybees must interact with the thousands of species of native Australian animals that depend on harvesting nectar or pollen to make their living. These include ants, beetles, bees, butterflies, flies, moths, and wasps along with many birds and small mammals. It is a diverse lot with various beak, snout, proboscis, and body designs—differences that no doubt make some of them better-adapted pollinators than others.

Most of the native Australian bees are quite small relative to the large honeybees, and they are often active at different times of the day. All of these behavioral and morphological differences are bound to make a difference in the level and quality of pollinator services offered by these native and exotic animals to the Australian flowering plants. David Paton's examination of bottlebrushes *(Callistemon)* reveals that honeybees sometimes remove more than 90 percent of the floral resources produced by these plants. "Clearly," he says, "interactions between honey bees and at least some of the Australian biota are not trivial. . . . The amount of nectar removal from these 'bird blossoms' decreased in direct proportion to increases in honey bee floral visitation. There is, however, a limit to the amount of nectar the honey bees can steal away and sequester." Much of the nectar is secreted at night, which means that the "early birds" can get the nectar before the honeybees wake up. By midmorning, however, with the "call to nectar" danced into their tiny brains by the scout bees, the flowers are often swamped by efficiently recruited hordes of honeybee foragers. There is little surplus food left among the blossoms. Pollen and nectar, therefore, may be in short supply for both honeybees and the little native bees.

Australian "bugs" now have to share most of their floral pantries with the introduced European imports. The native bees are much smaller than *Apis* and have smaller energy demands per individual. Too, the little colletids forage during the heat of midday when honeybees are much less active. Honeybees usually forage at lower temperatures, which gives them first crack at the large amounts of nectar present during early morning hours. This foraging lifestyle does seem to give honeybees a distinct advantage in most competitive interactions, especially if nectar production peaks in the morning.

But is the consumption of pollen or nectar by honeybees really detrimental to the lives of native bees in Australia? Paton examined the trophic relationships between 100 species of Australian nectar-feeding birds—mostly honeyeaters in the family Meliphagidae—and honeybees. In southeastern Australia, the scarlet bottlebrush *(Callistemon rugulosus)* is visited heavily by the New Holland honeyeater

(*Phylidonyris novaehollandiae*), whose breeding is correlated with the bloom period of this plant. Paton found that the honeyeaters were losing up to 50 percent of the available nectar to honeybees. When few honeybees were present, the birds visited individual flowers 9.6 times per day; when the bee numbers increased, the bird visits dropped to only 3 per flower per day. When Paton increased the number of honeybees present by introducing ten hives next to a patch of bottlebrush, the dominant adult male birds expanded their feeding territories by displacing other birds indicating that the new honeybee interlopers were getting much of the share of sweets.

Paton also offers some interesting insights from the plant's perspective. Bottlebrush plants need cross-pollination to set appreciable levels of fruit. But honeybees working the flowers only struck the receptive stigmas 4 percent of the time they were studied. The larger honeyeaters, by contrast, hit the mark and deposited pollen grains more than half the time. Honeybees, moreover, rarely moved between adjacent plants. In short, they simply did not facilitate outcrossing to the extent that the birds did.

In recent years, a war has been raging in the western United States between cattlemen and wilderness advocates. The cattlemen claim that their stock make productive use of rangelands too dry, too stony, or too steep for agriculture or other uses, allowing their families to make an efficient living in remote areas. Wilderness advocates, on the other hand, claim that cattle are not native to the Americas, that they unfairly compete with wildlife, that they have diminished the diversity of life the wildlands were capable once of sustaining.

In fact, the honeybee has perhaps had as much impact as cattle on the structure of certain plant communities. The debate between park preservationists and migratory beekeepers in Australia has much the same fervor as that between Earth First!ers and the Wise Use movement in the United States. The lines were drawn long ago, and few naturalists today can comfortably walk the middle ground between these two opposite poles.

It may be apocalyptic to claim that someday beekeeping with either European or Africanized honeybees will be discussed in the United States and Mexico with as much emotion as cowboying is today, but that is indeed our prophecy. Honeybees are, after all, lilliputian livestock—

fuzzy herbivores with wings—that are just as capable of taming a landscape as any cow, sheep, or goat infestation. Their "grazing" on pollen and nectar simply goes unnoticed. They may buzz softly, but they pack a big ecological wallop when it comes to altering, perhaps forever, the potential mix of forages out there on the range, in the bush, in the outback or boonies.

*This scene of a crop-dusting airplane spewing deadly insecticides is a common one in agricultural lands in the United States and other countries. Often these chemical brews are complex mixes that kill honeybees, native bees, and other pollinators in addition to their targeted herbivorous crop pests. Growers and pest control applicators can work together by choosing less toxic materials—and applying them only when needed, when pollinators are less active, and when crops are not in bloom.*

# CHAPTER 11

# The Little Lives
# Keeping Crops Fruitful

### The Economics of Pollination

The Great Basin Desert does not immediately seem to be a place where you would go for a remedial lesson in the agricultural economics of pollination. At first look, it does not even seem to be a place well suited to agriculture at all. Its soils can make life difficult for plants, for they may be loaded with white salt crystals, black alkaline crusts, or sickly yellow sulfuric acids derived from shale outcrops. It is a landscape largely covered by gray, uninviting shrubs such as big sagebrush, shadscale, saltbush, and blackbrush. Summer daytime temperatures can climb past 110 degrees, and plummet to 20 below at night in the winter, and it never, ever rains enough. To a grain and forage farmer—or to a honeybee—it may seem as remote as the land of Oz from the agrarian ideal found in fertile breadbaskets like Kansas.

The Great Basin is farmed nonetheless. Its leguminous forage and grain crops are cultivated by a mix of ingenious, hardworking Mormons, Basques, Japanese, Scandinavians, and other ethnic farmers. For irrigation, they have developed shallow wells, or placed their fields on sandier soils downstream from the edges of silty, saline seeps. Some draw upon natural springs from which they gravity-feed water through irrigation canals.

It is in these settings—near seeps and springs marked by crunchy calcium and magnesium crusts—that Great Basin farmers found for themselves a pleasant surprise. A native ground-nesting bee there, *Nomia melanderi,* frequents the moist alkaline crust soils surrounding natural seepages. And whenever farmers planted alfalfa, sweet clover, spearmint, or peppermint, the bees soon began to increase in abundance due to the increased floral "bee pasture" offered by these large-scale plantings. During the first half of the twentieth century, central Utah had become the nation's principal alfalfa seed-growing area, with yields of 300 to 600 pounds of alfalfa seed per acre. After modernization—and with the efficient alkali and leafcutter bees as managed pollinators— seed yields today are routinely over 1,000 pounds of clean alfalfa seed per acre and often achieve the staggering level of 2,400 pounds per acre. That's a lot of alfalfa sprouts in the supermarket. By 1940, it was real- ized that alkali bees—not honeybees—were responsible for pollinating more than half the alfalfa flowers found in a field. Each female alkali bee was personally responsible for producing a fifth to a third of a pound of alfalfa seed over the growing season!

While honeybees can pollinate alfalfa flowers, they have no predilec- tion for "tripping" these complex legume blossoms—that is, they seldom "unhinge" the keel and wing petals held under tension to release the sta- mens and stigma protected within. The alkali bee, though, is a master at this maneuver, busily tripping more flowers per legume than an indi- vidual plant can mature as ripe fruits. Historically, this bee naturally de- pended on native leguminous herbs such as locoweed and lotus scattered between the gray and greasy wind-pollinated shrubs of the Great Basin high desert. A female alkali bee can pollinate 2,000 legume flowers in a day, and as many as 25,000 flowers over her lifetime.

By the late 1940s, alfalfa seed growers realized that these native bees— easy to spot with their pearly iridescent abdominal stripes—could be se- lectively managed to the farmer's benefit. If as many as 2,500 female bees could be encouraged to congregate on a single acre of fieldside habitat, the pollination needs of a hundred acres of alfalfa or mint would be com-

*A female alkali bee* (Nomia melanderi) *is about to pollinate an alfalfa blossom* (Medicago sativa) *in the Pacific Northwest of the United States. This bee is native to the region, nests gregariously, and was the first ground-nesting bee to be managed as a pollinator of commercial alfalfa plantings. A recently visited "tripped" flower can be seen just below the bee. Managed "bee beds" have been decimated, however, by aerial pesticide applications.*

pletely met. Nearby farmers became impressed with the alkali bee's pollination abilities. The first artificially constructed and maintained bee bed was established in the early 1950s outside Boise, Idaho, on the Collard family farm, assisted by scientists at Oregon State University. Ned Bohart first, then followed later by W. P. Stephen and Phil Torchio, helped farmers create nesting beds for the bees in Oregon, California, Idaho, and Utah between 1956 and 1960, adapting techniques to the conditions of each region. Farmers soon learned to transplant and to propagate these native bee colonies to managed artificial bee beds adjacent to not only their alfalfa but also their mint, onion, and celery plantings.

And that is how an amazing agricultural innovation evolved. Rather than relying on the well-rutted routines of managing exotic social honeybees, Great Basin farmers improvised the means of taking a "plug" or "divot" of bee-bearing soil from a native congregation to use in salting a fieldside area to establish a new nest site for these solitary but aggregating bees. By the 1960s, enthusiastic farmers had established co-operatives to build and protect such nesting sites. In the history of the use of honeybees and other bees in U.S. agriculture, it has been rare to see the propagation, development—we hate to use the term "domesti-cation"—and commercial utilization of a new pollinator brought up from the ranks of native bees. Such was the case, however, with the al-kali bee on alfalfa acreages.

S TEVE REMEMBERS:
Over a decade ago, I encountered one of these managed alkali bee beds in eastern Washington, holding perhaps a few thousand bees. It was a quarter mile long, forming a 25 to 30 foot wide strip along the edge of an alfalfa field. By taking a shovel and digging down through the bright white salt surface crust to the moist and sandy soil below, farmers discovered that the alkali bees nest shallowly, only 6 to 8 inches deep. Later they learned that by bulldozing out a trough perhaps 2 or 3 feet deep and lining the bottom with an impermeable layer of plastic or bentonite clay, they could begin to create a semblance of an alkaline playa where female alkali bees nest. By piling large rocks, gravel, cheesecloth, sand, topsoil, then finely ground rock salt on top of the impermeable layer, they formed a matrix suitable for alkali bee burrows. Next they placed concrete standpipes 50 to 75 feet apart so that the entire bee bed can be inundated and moisture can percolate upward, just as it does in natural alkali seeps. Then they took bee blocks—plugs loaded with bee pupae in their natal cells from natural nest aggregations—and used them to salt the new beds. Finally, they placed shade shelters over the bee beds in some regions by covering wooden frames with war-surplus camou-flage. As Phil Torchio told me, "these shades worked like a charm.

Temperatures at midday are reduced by 15 to 17 degrees under the camouflage." Torchio also had to impress on farmers the need to keep sufficient pollen and nectar forage around for the bees on adjacent wildlands or irrigated pastures, or subsequent generations of alkali bees would dwindle. Where the farmers took heed of his admonitions, the bees survived.

For several decades, alkali bees were the key to alfalfa seed production, ensuring much higher yields in the American West for seed that was then distributed worldwide. This seed alone, destined for vegetable sprouts in addition to seed stocks for replanting, was valued at $115 million by 1990. Why? Because alfalfa hay is the premier forage crop in North America, surpassed only in acreage by soybeans as a field crop. By 1990, alfalfa hay was worth nearly $5 billion a year, and combined with the value of derived products, contributed over $12 billion a year to the U.S. economy. While hardly anyone outside of the Great Basin was aware of it, native bees were responsible for the bulk of the seeding success of the most important hay crop used for livestock fodder in the western United States. Nearly 85 percent of the alfalfa seed crop in the United States was produced on land serviced by alkali bees and leaf-cutter bees *(Megachile rotundata)*—an area comprising only 15 percent of the land dedicated to alfalfa seed production.

Then trouble hit the homelands of the alkali bees in the Great Basin and the Pacific Northwest, at a time when they had virtually no one to champion their cause and protect them. Ned Bohart from the USDA-ARS laboratory in Logan, Utah, remembers the general sequence of events in this way: "The older farmers in central Utah recall that in the 1920s and 1930s there were many [natural] alkali bee nesting sites with populations running into the millions. Although they did not realize it, these bees were responsible for the high yields of alfalfa seed that made Utah one of the nation's principal producers at that time." Bohart continues:

Then the increasing demand for seed induced many farmers to plough their sites, even though the soil was highly alkaline. Reduction in the population was immediate. I recall hearing a farmer from Delta, Utah, describe huge flocks of seagulls that followed his plow to feed on the overwintering bee larvae which were strewn behind him like popcorn as he plowed. In the 1940s, much of the area around Delta had been drained and plowed.... In the 1950s, and early 1960s, dieldrin and parathion, which are highly toxic to alkali bees, were used extensively on alfalfa crops to control lygus bugs. As a result, the size and number of alkali bee nesting sites dwindled, and the bees were no longer important as pollinators except in peripheral areas where they partially escaped the hazards.

Insecticide spraying on range and farmlands killed an enormous percentage of the naturally occurring alkali bees in that state. In fact, the alkali bee was found to be somewhat more susceptible than the honeybee to most insecticides. In 1973 alone, inadvertent pesticide poisoning of alkali bee nesting areas resulted in $287,000 worth of lost alfalfa revenue from seed production in Washington state. Elsewhere the cooperatives that formed to protect alkali bee populations entered into covenants to control the use of insecticides within range of their alfalfa fields. Until recently, in Idaho, Oregon, and parts of Washington these covenants allowed the persistence of the farmer/bee mutualism.

Nevertheless, even the alkali bees protected in artificial nest sites have suffered declines due to yeast and mold spoilage of their underground food stores following heavy summer rains beginning in the early 1960s. Although they remain marginally viable in Washington and parts of Utah, their numbers are nothing compared to what they once were. Belatedly—perhaps too late—they have found a few vocal advocates to protect them from further harm.

The demise of a native bee hardly makes the front page or a spot on the six o'clock news. But now the American honeybee industry is faltering. It is declining rapidly, too, due to diverse threats—from two mite species, from fungal, bacterial, and viral diseases, and from a host of pesticides, herbicides, and economic threats including declining price supports in the face of cheap imported honey from China and Mexico. At the same time, Africanized bees are expanding their occupied range and threaten this vital pollination service. While the assaults on managed European honeybees do make headline news, few reporters ask the critical question: If honeybees continue along their present declining trajectory, what bees or other pollinators will take up the slack in providing essential pollination services for our vast commercial and home agricultural plantings?

Two talented brothers—economist Lawrence Southwick and the late bee biologist Edward Southwick, both of New York—once asked, and attempted to answer, that very question. They assume that in the northern United States, 50 percent of the managed European honeybee population may soon be lost due to tracheal and Varroa mites along with other parasites and diseases. In addition, there may also be a complete abandonment of managed European honeybee colonies in the southern tier of U.S. states as the encroaching Africanized subspecies arrive and increase beekeepers' liabilities. If these projected losses materialize, as the Southwick brothers suggest, then one-fifth of the alfalfa crop may be lost. And if alkali bees and other species such as alfalfa leafcutter bees are not further recruited to fill the void left by declining honeybee colonies, there could be a 70 percent crop reduction in the alfalfa fields of the United States.

If these projected crop losses are translated into society's economic losses in terms of reduced availability of alfalfa, or higher prices, the cost of diminished honeybees with no wild pollinators to replace them would amount to $315 million a year. If, however, native pollinators were somehow revived and managed to fill the niche vacated by commercial and feral honeybee colonies, the demise of this one species of pollinator alone would cost U.S. consumers "only" $40.5 million each year. As astonishing as it may sound, the propagation and management of alkali bees, leafcutter bees, and their other native cousins to replace honeybees could result in the saving of $275 million dollars to U.S. alfalfa producers, growers, and consumers every year.

When the Southwick brothers extended their analysis from alfalfa pollination to include over 60 other U.S. crops, the unsung economic contributions of native pollinators, especially bees, suddenly became evident. If honeybee colonies near farmlands were to decline as outlined above—a 50 percent decline in the north and a probable 100 percent decline in the south—there would no doubt be annual accumulated losses in the billions of dollars to the American economy. The totals are fairly shocking: if no native pollinating bees replaced the alien honeybees in providing vital pollination services, the annual losses could rise as high as $5.7 billion. Even in the fedspeak of billions and billions of dollars, this is not small change.

Alternative pollinators—wild ones like nectar-feeding bats, hawk-moths, butterflies, solitary bees, and nitidulid beetles, along with managed ones like alkali and alfalfa leafcutter bees—could reduce such projected losses to only $1.6 billion if they were properly cared for around farmlands. In essence, the Southwicks suggest that the potential value of native pollinators to the U.S. agroeconomy could be on the order of at least $4.1 billion per year!

Many Americans don't see their lives as being inextricably tied to the fate of alfalfa. They may not realize just how many other crops are fundamental to their current earnings and lifestyle, crops that are primarily pollinated by wild animals not under the care and stewardship of human farmers and others. As USDA bee biologists pointed out in 1983, two-thirds of the U.S.-produced commodities that benefit from insects take the form of alfalfa and its products.

Robert and Christine Prescott-Allen have surveyed 60 agricultural crops critical to the North American economy for their relative dependence on wild and semimanaged pollinators. They determined that seven crops currently worth about $1.25 billion annually in the United States are pollinated primarily by wild insects. These include cashews, squash, mangos, cardamon, cacao (chocolate), cranberries, and highbush blueberries. The other North American blueberry, the lowbush species—historically pollinated by at least 190 different kinds of native bees throughout its range—is now faced with insufficient rates of pollinator visitation within certain areas where it is produced. It appears that overzealous weed control by blueberry managers has resulted in the loss of habitat and forage for the native pollinators on which the lowbush blueberry used to depend. The production of another 18 major crops depends to some extent on wild pollinators. In addition, wild animals are critical to the production of seeds used in propagating another 19 crops in farmers' fields: onions, carrots, kapok, sunflowers, strawberries, cinnamon, clovers, figs, and coconuts are among those in this category.

In 1976, Samuel McGregor of the USDA-ARS in Tucson estimated that, in all, at least 150 major crops relied to some extent on wild pollinators. In a more recent survey, Dave Roubik documented that at least 800 cultivated plant species rely on wild bees and other insects for pollination. Roubik also tallied the breeding systems for some 1,330 cultivated crop species worldwide. Although there are surprising gaps in our knowledge about reproduction, pollination requirements, and just who is doing the pollinating for many of the most widespread crop plants, this compilation presents data that are far-reaching in nature. He suggests, for example, that about 73 percent of our cultivars are pollinated at least partially by a variety of bees, that 19 percent utilize flies to move their pollen, 6.5 percent use bats, 5 percent use wasps, 5 percent use beetles, 4 percent use birds, and 4 percent use butterflies and moths. Once again, our dependence on bees, other insects, and other pollinating animals is clearly demonstrated. Eliminating these pollen-moving creatures would take the food right out of our mouths.

A surprising spectrum of native invertebrates pollinates crop plants: midges, blowflies, soldier flies, syrphid flower flies, hawkmoths, nitidulid beetles, checkered beetles, ladybird beetles, fig wasps, sphecid and vespid wasps, sweat bees, squash and gourd bees, carpenter bees, andrenid bees, bumblebees, mason bees, leafcutter bees, and cactus-loving bees. Other crops are pollinated by a range of wild vertebrates, from nectar-feeding bats and opossums to perching birds such as orchard orioles, yellow-winged tanagers, singing blackbirds, brown jays, and golden-fronted woodpeckers. If such a diversity of wild pollinators is essential to the health and wealth of our agriculture—and that is an inescapable fact—why do so few people in the United States and elsewhere recognize that native animals are making significant contributions to crop ecosystems in general or to pollination in particular?

One reason is that the USDA and other compilers of agricultural statistics keep figures only on honeybees, and not other pollinators. To some extent, this is logistically reasonable, for it is far easier to tally managed honeybee colonies than it is to count wild beetles on custard apples in Florida and Puerto Rico. And so we know that by 1988 there were over 3.4 million honeybee colonies managed by 150,000 beekeepers, most of them hobbyists with fewer than 25 hives. Of the other beekeepers, 10,000 of them were part-time and 1,600 full-time commercial producers who together managed over 99 percent of all colonies. By 1994, the number of colonies kept by full-timers and part-timers had dropped from 3.2 million to 2.8 million. At least a million of these colonies were moved around and rented to farmers to provide crop pollination services during flowering seasons, so that 2 million colonies were more or less permanently rented out on farmlands during various times of the year.

And yet, even by USDA's own admission, honeybees currently provide only four-fifths of all insect pollination services received by cultivated crops in this country. Many critics say even this estimate is unusually high, that much of the pollination attributed to honeybees is really carried out by other floral visitors. (Our own surveys suggest that honeybees have been confirmed as the dominant pollinators for only 15 percent of the world's crops.) Nevertheless, that even one-fifth of all crop pollination is simply written off as done by "other unknown pollinators" is both astounding and dismaying.

Although the familiar honeybee, *Apis mellifera,* has spread across the face of the planet, it is not a panacea for pollinating our crops and especially not wildflowers. There are disadvantages to honeybees being so efficient at finding and removing floral resources before other pollinators find them. Moreover, they pack their pollen wet with nectar and saliva—thus rendering it inviable, or at the very least not so likely to be rubbed off and donated to a waiting stigma. In the words of James Thomson of the State University of New York at Stony Brook, this makes honeybees "ugly" pollinators, since they can actually undermine the activities of more allegiant native pollinators. Honeybees seem almost preadapted to living in naturally disturbed habitats or those disrupted by human events. This makes them somewhat "weedy," almost camp followers of man.

Simply put: honeybees are not a foolproof answer to providing our crop diversity with pollination services. They cannot be swapped interchangeably one-for-one with native bees or other pollinators. The fate of pollination in our rural communities, following the introduction of these social aliens, will never be the same again. It is a sobering thought indeed that as more and more of the globe is altered, the once ubiquitous honeybees are likely to pay pollinating visits to just a small fraction of the blossoms requiring pollination in the vast floral landscape.

What agricultural statistics do tell us is that the honeybee industry's ability to provide adequate pollination services has been declining for decades. And it is certain to decline even more in the coming years. The number of honeybee colonies in the United States peaked in 1947 at 5.9 million. The postwar rise in the use of organochlorine pesticides contributed to a 43 percent drop in U.S. honeybee colonies—from 5.9 million to only 4.1 million in 1972 and only 2.6 million in 1995. Even with the switch to organophosphate pesticides and better management of colonies to avoid pesticide poisoning, the industry has never fully recovered. Although official statistics define "colonies" by a different criterion these days, 3.4 million colonies is the highest annual estimate that has been recorded since 1947.

When we look at the factors currently limiting the honeybee's ability to provide pollination services, we find that several of the threats may be irresolvable. The honeybee tracheal mite, infecting the respiratory system of adult bees, was found for the first time in the United States in 1984, but it now is present in most states. In many states, beekeepers have reported colony losses as high as 50 percent due to this pest. If that were not enough, an external parasite known as the *Varroa* mite arrived in the United States three years later and now inflicts bee colonies in over 30 states. Fungal, protozoan, and bacterial diseases continue to afflict many colonies, as well, and the treatment only suppresses the infections without eliminating them. *Nosema* disease afflicts 60 percent of the colonies in the United States; foulbrood diseases affect another 2 percent. Together, these various pests and diseases result in losses of $192 million to beekeepers every year—not including reduced crop values due to diminished pollination services. The damages, moreover, are not limited to bee colonies in the field. Bee combs stored in warehouses are constantly under attack by gallerine moths whose caterpillars tunnel through the stored frames, greedily devouring the beeswax itself.

The bee industry was already crippled when Africanized bees decided to take up permanent residency in the continental United States in 1990, after more than a decade of isolated visits. Since then, they have helped carry mites into wildland and cropland populations of honeybees. In one area of Arizona, this has contributed to an 85 percent loss of feral honeybees in the last decade. Up to 80 percent of U.S. beekeepers may have to abandon their hives once Africanized bees arrive in their area. The one-time loss of 1.6 million honeybee colonies will reduce beekeepers' income by $160 million—but again, this estimate does not include the value of crops pollinated by honeybees. The value of that pollination service is estimated to be 50 to 60 times higher than the value of honey, wax, pollen, and other hive products.

The American honeybee industry has been changing rapidly anyway. Cheap honey imported largely from China, but also from vast apiaries in the state of Yucatan in Mexico, has been invading the U.S. market. Price supports for honey are being knocked out; the removal of subsidies dramatically affects the economics of managing bee colonies. Whether such agricultural subsidies are politically or philosophically right will not be debated here. Regardless of your stand on the matter, their loss has hurt the industry at a time when the country should be emphasizing, not deemphasizing, the importance of all pollinators. Clearly, as one recent USDA report has admitted, the honeybee industry is in a state of unprecendented crisis today. Ecologist James Thomson's response to this crisis is the same as that of many other pollination ecologists: "As insecticide use and habitat destruction continue to decimate wild bee populations, the need for managed pollinators is becoming more acute, and the value of honeybees as pollinators far outweighs their value as honey producers."

The recent decline of the honeybee industry is not necessarily irreversible. Nevertheless, it suggests that society would be wise to invest in the conservation and management of alternative pollinators, especially the numerous native bees which are excellent pollinators. Accordingly, Canadian pollination ecologist Peter Kevan has argued "for recognition of the limitations of honey bees and for expanded roles for native species in the commercial pollination of cultivated crops." In view of these trends, it is ironic that most research teams working with alternative pollinators have suffered staff reductions since 1990. Three-quarters of them expect further declines in team support. Following the first international workshop on non-*Apis* bees and their roles as crop pollinators, convened in Logan, Utah, in 1992, researchers were outspoken about the lack of public investment in obtaining the basic biological and cultural facts to help diversify pollinators serving agricultural crops and wildlands. As James Thomson observed:

> Certainly the participants at Logan collectively expressed a distinctive "attitude." This attitude embodied several elements: a long-standing frustration that honeybee studies and management programs receive disproportionate funding while the other 30,000 or so species of bees receive short shrift; an almost conspiratorial pleasure to be gathered in an "alternative" conference with like-minded researchers; a righteous sense that workers on non-*Apis* bees are doing work of considerable scientific and commercial value; and a cautious optimism that an avant-garde pollination biology may be forming.

Yet even if there were not impending honeybee, native bee, and pollination crises in the United States, there would be a number of valid reasons for promoting native pollinators for many cultivated crop plants. The benefits of native pollinators—even when honeybees are present—can be no better demonstrated than in the case of squashes, pumpkins, and gourds in the arid southwestern deserts of Arizona and northern Mexico. In the late 1970s Vince Tepedino studied squash pollination at the Greenville Farm in Utah, where the solitary gourd bee, *Peponapis pruinosa,* had been present since at least 1953. Honeybees were also common in the area. Between the two bees, one native and one introduced, 91 percent of all open-pollinated female squash blossoms set fruit. Tepedino sought to find out which of the bees was most responsible for that extremely high rate of fruit set.

Although his initial observations suggested that a single visit by either bee to a female squash blossom had the same probability of resulting in a mature squash, Tepedino kept working. He discovered that honeybees had a decided preference for visiting pistillate (female) flowers for nectar, whereas the native gourd bee preferred staminate (male) flowers, where they actively collected pollen. Moreover, the male native bees frequently encountered receptive female bees in the male flowers, where mating often occurred. More important, both sexes of native gourd bees were more adept than the honeybee at foraging from squash flowers. By the time honeybees took flight and arrived in open squash blossoms in the morning, the earlier-flying gourd bees had already pollinated them.

Our studies of wild gourds in the Sonoran Desert of Arizona and Mexico highlight the behavioral differences between native bees and honeybees as pollinators. The native gourd bees and carpenter bees were clearly the more reliable in harvesting and then depositing large quantities of pollen on receptive stigmas. We determined that to achieve full seed set for each female flower on a gourd or squash plant, it would take an average of 3.3 visits per honeybee, but as few as 1.3 visits per *Xenoglossa* squash bee and 1.1 visits per carpenter bee.

By other measures, too, all native squash and gourd bees are more effective than other gourd pollinators, including carpenter bees. They make more visits to gourd flowers than do other bees, and they time their visits so they arrive when pollen and nectar are most available. Not only are they faithful to squashes and gourds, but they are strong fliers that frequently move pollen between far-flung plants of the same species, thereby promoting genetic diversity in the population.

In fact, we documented a remarkable level of gourd bee fidelity to squashes and gourds—our bees hardly visited any plants other than those in the genus *Cucurbita*. The bees would often mate inside squash or gourd blossoms. Sometimes, their mating activity would simultaneously leave pollen on the stigmas of the blossoms. Pollen-covered gourd bees spend lengthy periods of time perching, grooming, and waiting on the massive gourd stigmas—behavior never observed among honeybees, nor among other native bees for that matter. These are secretive little lives that feed us.

And when we mistreat these little lives, the repercussions may suddenly cascade in from unknown directions. Look what happened when the Canadian government substituted the biodegradable organophosphate known as Fenitrothion when DDT fell into disfavor. Both were applied to literally tens of thousands of acres of coniferous forest in an attempt to control spruce–budworm infestations in New Brunswick. Peter Kevan has documented how Fenitrothion was indirectly responsible for dramatic drops in commercial lowbush blueberry production in adjacent areas. With the advent of widespread Fenitrothion aerial sprayings, blueberry crop yields in New Brunswick plummeted from

5.5 million pounds in 1969 to only 1.5 million pounds in 1970. Over a four-year period, New Brunswick's losses averaged a million pounds per season whereas blueberry yields in nearby Nova Scotia remained relatively stable.

Over the following decade, Kevan interpreted the yield losses in New Brunswick blueberry plantations from a variety of ecological perspectives. He calls the following scenario his "blueberry pie" model of the relationship between crop, pollinators, and other animals. From interviews in the field, he learned that "blueberry farmers blame the whole problem on the forestry industry, and in particular on the spruce budworm control program's use of Fenitrothion. The contention of the blueberry farmers has been demonstrated as true; pollinators are killed by Fenitrothion on blueberry fields." In fact, this organophosphate, which is closely related to wartime nerve gases, is far more toxic than DDT to native solitary bees and social bumblebees. In addition, the timing of spruce budworm spraying coincides exactly with the period of maximum blueberry flowering. The bumblebees and sweat bees that normally pollinate blueberries were noticeably scarce during the 1970 season on fields adjacent to forests where spraying had been undertaken to control the moths. Accordingly, berry yields dropped until honeybee colonies were introduced by commercial beekeepers and growers to make up for the lack of wild bee pollinators—and only recovered fully when the local use of Fenitrothion was eventually abandoned several years later. Kevan and his coworkers have documented that in some cases a return to "normal" population levels for bumblebees and other bees has taken as long as eight years.

At the same time, ironically, there was also unprecedented blueberry damage by native songbirds in 1970—damage so severe that there was a public outcry against blueberry growers who were shooting the robins, starlings, cowbirds, and waxwings that were denuding their berry fields. What was going on? Kevan believes that Fenitrothion was insidiously knocking back the numerous parasitic wasps and disease vectors (avian malaria carried by mosquitoes) that had been keeping the blueberry-gorging bird populations in check. Environmental disruptions, he says, have removed some of the factors that limit the bird populations: "For the birds, the factors could be predators and/or disease. The flocks of starving but otherwise naturally healthy birds would invade the crops to feed. . . . Thus, the blueberry yield losses can be attributed to two causes, both resulting from the use of Fenitrothion: reduced pollination by native bees and increased fruit damage by large bird populations, . . . resulting in losses of millions of dollars."

But careless use of agrichemicals is not the only threat to pollination and thus the stability of our food supply. Other unprecedented threats have been ushered in by the genetic engineering of ultramodern commercial crops, a high stakes game played out by large multinational companies here and abroad. When "biotechnologists" transfer a single gene that confers an advantage—such as herbicide resistance, insect resistance, or disease resistance (which make up the vast majority of today's transgenic plants)—into crops that require cross-fertilization, they may inadvertently allow these genes to escape into cross-compatible weedy relatives. Many scientists have wondered if this will, in fact, create superweeds. Norman Ellstrand of the University of California at Riverside and other scientists have conducted experiments with innocuous genetic markers to find out whether crops do, on occasion, mate with their weedy relatives under normal agricultural conditions. A growing number of such studies have confirmed that they do. This accidental but equally efficient gene transfer easily occurs where bees and other insect pollinators are wide-ranging and in places where weed populations are as distant as several hundred meters. Ellstrand further predicts that small, fragmented populations of weedy habitats near agricultural fields are even more likely to mate with large crop stands when disturbed by human activity.

The uncontrolled transfer of advantageous crop genes into weedy populations is not unique to the products of biotechnology. The precedent has already been played out with traditional crops and their weedy relations. Human hardship, in terms of reduced crop yield, has already occurred after weedy relatives of rice, sugar beet, radish, sorghum, rye, pearl millet, and probably several others have become more aggressive from picking up new genes from nearby crops.

In some cases that we have documented in Arizona and Sonora, Mexico—as with wild chiles (*Capsicum annuum*)—gene exchange may be regarded as positive. But in other cases, bitter squashes and quick-growing weedy radishes will win no blue ribbons at county fairs. If Ellstrand is right, then we must pause to consider that genetic engineering may pose significant risks where plant/pollinator relationships are already out of kilter due to disturbed landscapes—or where we know almost nothing about who pollinates what to begin with.

It is evident, then, that the little pollinator go-betweens linking the planet's flowering plants are being sadly mistreated by us, *Homo sapiens,* with our ceaseless destructive changes. At the very same time we are beginning to recognize the inestimable value, often indirect, of all wild pollinators to our food supply, we are also beginning to fathom how profoundly we have disrupted their relationships with plants.

*Pollinator gardens and other attractants for bees, birds, and butterflies are becoming very popular worldwide. Here, in a backyard vegetable and flower garden, are hummingbird and butterfly feeders, flowering plants providing nectar, pollen, and leaves (for caterpillars), and simple-to-construct homes (cans of paper soda straws or drilled wooden nesting blocks for leafcutter and mason bees) for pollinators. These techniques not only ensure a bountiful harvest but increase the local pollinators, many at risk due to habitat loss or pesticide abuse.*

# CHAPTER 12

# Cultivating Lasting Relationships

## Pollinator Gardens and Ecological Restoration

G ARY REMEMBERS:

Dawn came with a certain deep thrumming noise close to the screened-in porch where I had been sleeping. A pale lemon light diffused through a scattered bank of clouds above the horizon. It was spring, when black-chinned hummingbirds arrive to celebrate the blood-colored flowers waggling in the breeze on gangly ocotillo branches. High on the gently rolling ridges west of Tucson, thousands of wild-armed ocotillos were painting the desert grasslands a fiery red. The ocotillos just outside my porch, however, had been planted—along with a variety of other herbs, shrubs, and succulents—to attract particular moths, butterflies, solitary bees, and birds.

As the morning progressed, I would silently tally the visitors to the flowers in my backyard desert oasis. These flowers were all native to the Sonoran Desert region, but were arranged in ways that multiplied the pleasures of my private pollinator garden.

I disappeared inside the house and put the teapot on. By the time I had dressed, some chamomile flower tea was ready to accompany me on my early morning stroll around the yard. My first steps out the door dropped me down to the lowest depression in the yard, where the night air still hung sweet and low over the squash and gourd garden. There, just as I imagined they would be, early-rising squash and gourd bees were already pollinating the yellow-orange blossoms on the vines sprawling over the ground. A low buzzing sound was heard in a wild coyote gourd blossom. It confirmed that the gourd bees had already been active, navigating between blossoms, even though the sun was not yet high enough to touch this part of the yard and tease the blossoms into fully opening.

With a flashlight in one hand and a coffee mug in the other, I caught a glimpse of a solitary bee moving around the golden anthers of a male flower. Nearby, I spotted a female gourd bee landing on the ground to bring a massive load of orange pollen into her nest in the dew-covered soil beneath the pungent vines.

By the time sunlight began its cascade into the yard, I could at last see the black-chinned and Costa's hummingbirds moving between sprays of ocotillo blossoms, chuparosa tubes, and handmade, crimson-red sugar feeders hanging from the porch rafters. One Costa's humming-bird hovered at the feeder briefly before a black, orange, and white hooded oriole scared it away by flying to an orange-rimmed feeder nearby.

The oriole warbled a lovely territorial song from its perch by the feeder. The hummer noisily careened all the way across the yard, to land on the ramada shelter I had built out of more ocotillo. The Costa's male rested on the roof of this rustic bower, then began probing the flowers still blossoming from the pruned ocotillo branches I had laid across the vigas of the ramada. It was an odd sight: a Costa's humming-bird pollinating a piece of still-living architecture, a branch pruned months before, then strewn across the ramada roof.

It wasn't long before other bees—both social and solitary ones—were abuzz on the birds-of-paradise and mesquites surrounding the patio. The long, brilliant red filaments of the birds-of-paradise looked as though they would attract hummingbirds, but carpenter and honeybees were their more common visitors, sidestepping the anthers and robbing nectar. The carpenter bees were especially adept at floral burglary, for they had stiletto-like mouthparts that could slit the throat of any tubular flower in their way.

By this time I had finished my tea, and with garden hose in hand was watering my butterfly and bee garden. This early in the spring, however, only black swallowtails had appeared with any frequency. While watering the milkweeds and pipevines—larval host plants for queens and pipevine swallowtails, respectively—I searched their stalks and foliage for signs of the first young larvae. Although I didn't sight any this particular day, I still found the search for larvae as exciting—and sometimes as frustrating—as netting and then releasing the first mature butterflies in the yard.

As I moved the soaker hose from plant to plant, I tried to remember what kind of butterfly or moth each herb or shrub typically serves as a caterpillar chef's salad. My passionflower vines have made great larval food plants for the *Hegesia* fritillary and the zebra butterfly, both of which are fairly rare in the United States beyond the Arizona-Mexico border. The moist ground around verbenas and lantanas kept a good flock of *Melinus* hairstreaks in residence, while checkered whites seemed to like the jackass clovers and tansy mustards at the yard's edge. A *Buddleia* "butterfly bush" from the Chihuahuan Desert in New Mexico created a cluster of exotic cabbage white butterflies, as well as the more delicate orange tips, a native in the pierid family.

Moving closer to the orchard, I checked the occupancy status of the wooden nesting blocks for orchard mason bees, managed ones I had set out early in February. The bees must have spotted the For Sale sign, for this is prime bee real estate. On an unseasonably warm day in mid-February, one of these *Osmia* bees—probably a male—burrowed his way out from behind a thick mud barrier, the first guy on the block to emerge. None of his neighbors felt ready to follow him out, though, and he soon disappeared as a cold snap came on, bringing snow. After the snow had melted, and warmth returned to the desert, the rest of the bees in the nest block emerged—they were mostly females—and began to pollinate the apples, apricots, and nectarines in the mini-orchard within the yard. The females brought load after load of nectar and pollen back to the drilled wooden nesting block for the eggs they would soon lay within the dark tunnels.

I then watered a mob of bat-pollinated century plants, a few towering moth-pollinated soapweed yuccas, a hawkmoth-loving jimsonweed, some wild chiles pollinated by various sweat bees, and a dozen or so other hummingbird plants, some of which also attracted honeybees. Beneath the giant canopies of mature mesquites, there were thousands of tiny solitary *(Perdita)* bees congregating; there were also a night-blooming cereus and a few evening primroses waiting there in the shadows for hawkmoths to build up numbers later in the season.

Of course, not all of these plants would bloom in the coming year. Nor did all of them form patches large enough to attract and sustain a resident population of pollinators. But with all the additional pollen and nectar available from forage plants in the surrounding wildlands, many pollinator species would come and go from the yard as they pleased. And perhaps that is how it should be: a seamless transition from wild to cultivated, and back again.

Such personal and public display gardens, of course, give considerable satisfaction to their stewards. But do such plantings have any true conservation value in the face of all the human disturbance disrupting the subtle and complex interactions between plants and pollinators on a global scale? In and of itself, a small pollinator garden for butterflies, hummingbirds, or bees cannot possibly conserve entire populations or species of threatened or endangered pollinators and plants. It can, however, remind gardeners, neighbors, and others of the primacy of the precious keystone relationships between them. As Stanwyn Shetler of the Smithsonian Institution once warned:

> One butterfly or one wildflower does not an ecosystem make. Nature's rich complement of butterflies and flowers and other organisms and their myriad, evolved relationships can only survive if the diversity of the natural habitat is preserved. [And yet] the butterfly garden is a wonderful window on the local environment. Like a light trap or a bird feeder, a butterfly garden lets you know what is in the territory or which way the ecological currents are blowing. It is a telling index of the character and well-being of the neighboring patches of nature.

To be sure, we are seldom moved to protect any life that we haven't first noticed and then grown to know intimately and to love. Such pollinator gardens can stave off what Xerces Society founder Bob Pyle calls "the extinction of experience": the loss of direct contact with wildlife that leads to a cycle of disaffection, apathy, irresponsibility, and sometimes outright contempt toward natural habitats. If backyard bugwatchers become engaged with the interactions they see in their gardens, they may become stronger advocates for keeping highway medians herbicide-free, for assuring that only biological control agents are used to manage pests on food crops, and for establishing corridors linking protected areas.

Of course, gardening constituencies in the United States are enormous, so the conservation advocates among them may already be quite numerous. As the National Gardening Association has learned through its surveys, close to 44 million Americans are involved in flower gardening, 30 million grow vegetables, and of the 26 million actively landscaping their yards, an increasing percentage are planting natives. Roughly half of all households in the United States include at least one person involved in these three outdoor activities, and there is great potential for engaging them, not only with plants, but with butterflies, bees, and other beneficial "bugs." Worldwide the number of people involved not just in recreational but day-to-day subsistence farming is astronomical. Further, the total combined value derived from all agricultural commodities around the globe adds up to a staggering $3 trillion.

Another response to the effectiveness of pollinator gardens relates to their context: if they are linked in some way with larger efforts toward ecological restoration, each experimental garden can become incredibly significant. It does not matter so much whether the initial garden is relatively small, so long as some techniques are learned that can eventually help to heal physically or chemically damaged or otherwise depleted lands. We are reminded of the Quixote Principle elaborated by our fellow Tucsonan Jim Corbett, founder of the Sanctuary Movement: "The social significance of a cultural breakthrough, as contrasted with a social movement, does not arise from its being done by vast masses of people, but from its being decisively done by someone."

If there is one person who clearly demonstrates how a "cultural breakthrough" such as pollinator gardening can lead to a "social move-ment" with profound conservation implications, it is Miriam Roth-schild. Born in 1908 to an eccentric family of British naturalists, Miriam Rothschild has made innovative contributions to the fields of ento-mology, chemistry, pharmacology, marine ecology, human rights, and, oddly, to public safety: she was the first person ever to install seat belts in a motor vehicle. Nevertheless, her most lasting contribution may be her initiation of "butterfly gardening" in the British Isles. Over the last two decades, her butterfly gardening has expanded as a means to restore the biologically diverse cultural landscape she calls "the flowering hay-field." Consider, in her own words, what she has been up against in that effort:

> In England we live in a green desert, where modern agricultural methods have bulldozed, drained and sprayed all the buttercups and daisies out of the grass, the poppies out of the arable crops, and pulled up the dog roses along the hedges to make life easier for the combine-harvester trundling along the horizon.
>
> I decided to bring medieval flowering hayfields back to the countryside. Recreating such a meadow, I was told by the scientists, would take a thou-sand years. I believe them—but have produced a remarkably good imita-tion within a decade. . . .
>
> Eventually I counted ninety wild species [of plants] in the field, [but] the right sort of insects were slower to catch on. But after about seven years, during which time I eagerly awaited their arrival every summer, the first Meadow Brown butterfly flapped out of the grass as the dogs chased one an-other in high spirits. Within a day or two I spotted a bright azure fragment of sky caught in a patch of yellow trefoil—the first Common blue butterfly had arrived. Now they are breeding in the meadow. . . . The ants have set-tled in too, and so have the bumblebees. Both are good pollinators in their own right.

Miriam Rothschild began her flowering hayfield by letting wild grass grow up over 2 acres of bowling greens and tennis courts at Ashton Wold, her family's estate. She then collected seeds from nearby forests, hedgerows, and meadows, sowing them directly into the hay-field or transplanting out more vulnerable species when their seedlings

*British hedgerows have traditionally provided shelter and places to feed for wildlife including many pollinators. Formerly under attack by uniform farming methods, they have been planted once again in many areas. In the European community, such set-aside programs are invaluable for protecting pollinators.*

were hardy enough to compete on their own. She put special effort into collecting plants that produced abundant nectar. These nectar plants would sustain the butterflies that she and her father have loved and studied. Nectar plants, she thinks, are the key to diversifying the pollinator fauna: "You can really abandon any romantic idea of creating a home for these angelic creatures," she says dryly, adding, "the best you can do is provide them with a good pub."

Once Rothschild had given the meadow some momentum on its way to restoration, she noticed a curious phenomenon: natural processes took over from there. Several other species—from bee orchids to fairy flax and quaking grass—arrived by aid of the wind or small animals. "Others," she hopes, "will follow suit."

Others have indeed followed suit, thanks to Miriam Rothschild's inspiration—and to the wildflower seed supply she began to grow on another 90-acre parcel. Her efforts provided the nursery grounds from which many other pollinator conservation efforts have grown. Bob Pyle has told us that until Miriam Rothschild came along, "nearly all butterfly gardens were postage stamp in size. Then Miriam Rothschild began reclaiming waste sites, open cast mining and reservoir sites, and covered reservoirs. It was just in time."

Until Rothschild's efforts began to be emulated, the local extinction rates for butterflies in some regions of England were ten times higher than those for vertebrates or vascular plants. As butterfly conservationist Tim New has noted, "conservation management evolved slowly in Britain, [through] years of trial and error." Rothschild gave it a kick in the seat—proposing the sowing of native nectar plants in the half million acres of private gardens in Great Britain, as well as the regreening of highway right-of-ways (verges) and landfills to increase the local abundance of rare butterflies. After seeing pilot projects by Michael Warren and other butterfly conservationists, the British Roadside Authority decided to make all of their medians and verges more wildlife-friendly. By resculpting the topography around a sanitary landfill in Essex, then sowing it with butterfly food plants, the new generation of conservation biologists has now demonstrated that rare butterflies can clearly benefit from the ecological restoration of their habitats.

More remarkably, British conservationists have begun to realize that maintaining traditional land management techniques is more important to butterfly conservation than simply setting aside protected areas. This realization did not come in a flash. The heath fritillary disappeared from numerous sites in Great Britain over the last three decades—even on two nature reserves specifically established in the 1960s to protect this rarity. Then conservation biologists reassessed what had gone wrong for the species they call *Mellicta athalia*. After studying its foraging behavior and physiology, they discovered that this fritillary is not so much dependent on protected "virgin" lands as it is on an ancient form of agroforestry that had been practiced in the British Isles for 6,000 years. That tradition of forestry management has largely been abandoned during the last few decades. It has been replaced by modern silvicultural techniques which open forest clearings that are too close together in space and time for the fritillary's liking. In some areas the management of protected reserves has been modified to emulate the ancient practice called "coppicing," a means for selectively cutting branches and regenerating trees. By coppicing oaks, hazels, and ashes—that is, encouraging thickets of small trees—conservationists have witnessed a comeback of blackthorn, bluebells, and primroses, and with them, the heath fritillary.

Recently they have translocated populations of this fritillary to isolated sites where the species formerly occurred. Once they reestablished traditional management practices there, they witnessed a rapid population growth that has been sustained for well over a decade now. In short, they discovered that this particular butterfly had come to rely on forest openings that traditional woodsmen had provided them for thousands of years. Heath fritillaries had become adapted, that is, to a cultural landscape.

Since that hard-won lesson with the heath fritillary has sunk in, British biologists have realized that other vulnerable pollinators are now confined to warm microclimates created over the centuries by human culture. In fact, these pollinators have not been found in any truly pristine habitats at any time over the last 300 years of record-keeping on the distributions of British invertebrates. Instead, they have taken to thin-soiled calcareous wetlands prehistorically excavated for peat, and to intensively coppiced woodlands where the tree canopy does not close up but stays in an arrested stage of ecological succession. In both of these open, sun-exposed habitat types, butterflies thrive where summer temperatures are several degrees higher than in nearby tree plantations or in lush, deep-soiled wetlands.

Encouragingly, in Britain, the conservation ethic for other invertebrates has caught on. Those green isles are a haven for a great number of bumblebee species, diversity that we only see elsewhere in the northeastern United States or Canada. Bumblebees favor, or at least tolerate, weather that is too cold or wet for honeybees and many native ground nesters. Great Britain is a nation of gardeners, astute naturalists, and conservationists. Perhaps this stems from a fine Victorian tradition that fostered Charles Darwin and friends absorbing local natural history at his country estate in Downe. Although it would probably amaze most

Americans, glued to their television sets, ambitious programs are under way in England for monitoring and conserving many invertebrates, including the fuzzy bumblebee genus *Bombus* and the parasitic cuckoo bee *Psithyrus*. The country has 22 bumblebee species, quite a few of which are rare, endangered, or locally extinct, due to the long history of agricultural land use and the disquieting recent destruction of vast numbers of ancient hedgrerows that once provided refuges to bees and other wildlife.

One of these new programs is Guardians of the Countryside, a major British conservation effort sponsored by Heinz Corporation through the World Wildlife Fund–U.K. to help protect wildlife and habitats at greatest risk. An ambitious and exciting program, Guardians distributed low-cost packets containing a bumblebee poster, a magnifying lens, an identification guide for all the bumblebees in Britain, and a packet of bumblebee wildflower seeds for planting. As part of Guardians and other invertebrate programs, a volunteer army of young and old naturalists now routinely reports to various local and regional centers the appearances—or notable absences—of common and rare bumblebees and other insects seasonally and yearly. This information is then cataloged into invaluable databases including modern "geographic information systems" so that land managers, citizens, indeed any interested person, can tap into this compendium of knowledge. We should be fostering such programs in the United States.

It recently dawned on many biologists that these British species, which take refuge in these culturally modified habitats, may be relicts from a warmer period that occurred more than 5,000 years ago when summer temperatures averaged 1 or 2 degrees Centigrade higher than they do today. Were it not for the human modifications that have kept these habitats warm and open over the last five millennia, many pollinators would surely have disappeared from the British Isles altogether. As Bob Pyle of the Xerces Society listened to his British colleagues tell one story after another like this, he came to understand that "virtually every endangered butterfly remaining in England has declined because of the abandonment or destruction of traditionally modified landscapes—hedgerows, hayfields, excavation sites, and coppiced woodlands." The same trend is evident for British bumblebees.

In North America, it is far harder to tell how pollinators were affected by the land management practices of Native Americans over the millennia, although their burning techniques likely favored population growth for Karner's blue and other rare butterflies. In the once extensive dunes along Lake Michigan's shoreline, generations of schoolchildren used to play hooky and start fires in the sandy grasslands and marshes, thereby maintaining a patchwork of different-aged vegetation. These patchworks—much like those shaped by Native American fire management—have kept habitats dynamic for several rare plants and insects already lost from other dune areas where static management has led to the local extirpation of the same species. Yet kids playing with fire is not the point. As another Lake Michigan dweller, Stephanie Mills, has written: "To be honorable, restoration activity must be inseparable from the prevention of further habitat destruction." And, of course, fire does not equally serve all plants and pollinators. Obviously, it is virtually useless as a management tool in an area too built over to sustain a habitat mosaic.

If dune management were to become dynamic again, would certain North American pollinators become more abundant? This question is being answered at the Antioch Dunes along the San Joaquin River in California, which we visited in an earlier chapter to assess the status of an endangered evening primrose. Some 20 acres in this short chain of dunes is the only remaining home of Lange's metalmark butterfly, *Apodemia mormo langei*. Since it was first listed as an endangered species in 1976, the Lange's metalmark population has been found to drop as low as 50 individuals during peak breeding season, a dangerously limited gene pool. For a while, this bright reddish-orange butterfly did not seem to have too bright a future—exotic weeds were outcompeting its primary host plant, a naked buckwheat known as *Eriogonum nudum* var. *auriculatum*. Since the caterpillars and adult females of this metalmark never stray far from the flowering buckwheats, it appeared that the butterflies would decline along with the buckwheat flowers.

But Mike Palmer, Stephanie Zador, and others with the U.S. Fish and Wildlife Service have witnessed a buckwheat revival on the low sand mounds they shaped in areas that had been formerly sandmined at

Antioch. This team has added 7,000 cubic yards of sand on 2 1/2 new acres of contoured dunes, planted out buckwheats, and achieved considerable success in seeing them come to seed. The metalmark butterfly population has responded accordingly. In fact, it has expanded to nearly 2,000 individuals seen on a single day in 1991. While the butterflies have been on a meteoric rise lately, it will require additional labor-intensive efforts to keep their host plants from suffering further competition with exotic weeds. Even so, by April 1995 the restoration project at Antioch Dunes could already be judged a success in providing more larval host plants and nectaries for the metalmarks.

Yet several questions remain unanswered with regard to the metalmark's use of restored vegetation. Not far from the artificial dunes lies an old vineyard that still contains a few grapevines as well as a proliferation of buckwheat. As Stephanie Zador pointed out to Gary, "the vineyard has tons of buckwheat but the metalmarks only visit its edges." She wonders whether there might still be some chemical residues in the vineyard soils—persistent pesticides or herbicides—that have kept the butterflies from foraging in this microhabitat.

It may not be possible to give her a definitive answer. What remains clear, however, is that the restoration of ecological conditions suitable for sustaining butterfly populations cannot simply stop with the planting of nectar and larval host plants in artificial dunes. As bioregionalist Stephanie Mills has warned: "The possibility of certain kinds of ecological restoration—establishing new patches of prairie plants, for instance—should not lure us into the delusion that ecosystems can be moved around like oriental rugs." The lesson from Antioch Dunes is telling. However vital it may be, the reintroduction of nectar and larval host plants is but a starting point for long-term ecological restoration.

The metalmark butterfly prefers certain microhabitats—perhaps due to their mix of herbs and openness, their blend of sunlight and protection from the winds—but rejects others that seem nearly identical in terms of their vegetation. To understand these subtle differences between one habitat and another "empty space" a hundred yards away, we must begin to see the world through the eyes of a bee or butterfly, to smell out its fragrances as a moth would do, and to taste the mix of sugars there as a hummingbird might taste them. Food, of course, is not the only requisite a pollinator must find in a particular place. In addition to pollen, and nectar, many pollinators require the local availability of standing water, as well as resins, saps, and gums to help in gluing together nesting materials. Nest sites—for a colonizing adult as well as its offspring—must be ensured. A suitable shelter may depend on the availability of nesting materials, holes or canopies, and on enough escape routes from potential predators. Each species of bee, hummingbird, or hawkmoth will meet these basic needs through different strategies.

Over the years we have experimented with designing various domiciles for native bees in order to understand their nesting requirements well enough to increase their diversity and abundance. We have found that some urban and agricultural landscapes suffer from a shortage of the holes required by pollinators—abandoned rodent burrows, hollow trees, old beetle galleries, cactus boots, miniature caves under cobbles and boulders. Our desert entomologist colleagues at the Sonoran Arthropod Studies Institute host an international conference called "Invertebrates in Captivity"—or "Bugs-in-Bondage," as Gary calls it— in Tucson each year. They work to devise ways of keeping insects captive for public exhibition in zoos and for teaching purposes in schools, and also to demonstrate the utility of using artificial nest sites and other materials in already damaged habitats where such resources have been severely depleted.

We have managed to increase the local density of native bees in our own backyards by using a variety of inexpensive materials as enticements. By packing old-fashioned "Sweetheart" brand paper straws into a milk carton, gluing them down, and affixing the carton to a branch within the shady canopy of a tree, we have "trap-nested" leafcutter and mason bees to encourage their repeated seasonal use of these artificial domiciles. We have also attracted bumblebee queens to nest near our gardens by leaving clay pots upside down on the ground near our tomato patch. We have coaxed other bumblebee queens to nest in the spring by partially burying a wooden box filled with upholsterer's cotton and drained by tubes and drilled holes. Native Sonoran bumblebees will readily colonize both kinds of shelters.

In shady protected sites adjacent to our gardens and orchards, we have hung drilled thick wooden boards and cans of straws or tied bundles of pithy stems. By drilling different-sized holes into blocks of these softwoods such as sugar pine and elderberry, an astonishing variety of native bees and wasps can be accommodated. Because natural nest sites are often limiting—bee real estate is hard to come by—drilled boards, styrofoam blocks with holes, grooved sheets of wood, and corrugated cardboard can help leafcutters along with other bees build up their populations and provide pollination for nearby gardens.

Of course, the small scale of most home gardens makes them easy to manage. The same pollinator-attracting techniques applied to the hayfield or restored sand dunes are less likely to do the trick. Nevertheless, alfalfa growers and collaborating scientists have established large artificial soil beds that attract more than a million alkali bees *(Nomia melanderi)* to acre after acre of moistened, salt-laden lands provided with occasional drain pipes. Such artificial beds can be created not only to service crops but to help pollinate reintroduced native plants as well.

It might seem a simple matter to establish reserves harboring interacting populations of diverse pollinators and their flowering plant hosts. This does not, however, seem to be the case. In fact, we have found few examples of pollinator and flower reserves set aside to preserve these delicate interactions. One place where the destruction of a diverse set of tropical flowering plants and their pollen-transporting go-betweens was halted, or at least forestalled, is at Lomas Barbudal. Nestled among picturesque volcanic hillsides in the dry deciduous forests of the northwestern cattle province of Guanacaste in Costa Rica, it preserves hundreds of floristically rich acres of wildlife habitat. Lomas Barbudal—literally "bearded hills," named for its dense but somewhat scruffy vegetative cover—is an island of protected land within an encroaching sea with sharp waves of chainsaws and machetes.

Ever since this area was opened up by the paved Pan American Highway, colonists and squatters have cut trails, paths, and dirt roadways through it, enabling them to hunt, fish, and log. Some of this colonization was haphazard and illegal; other programs were funded by multinational banks and actively sponsored by the Costa Rican government. Whatever the diverse causes, these once pristine forests and watersheds—part of the highly endangered and almost extinct remnants of the "dry tropical deciduous forest"—have been denuded to provide grasslands to raise beef cattle and horses. Slash-and-burn agriculture has been practiced for millennia here, but recently people have cut, burned and plowed up more forest than ever before. During the long dry season, the skies are choked and stained an ochre hue with the acrid smoke from dozens of fires set by ranchers and farmers. The caretaker and the resource managers of Lomas Barbudal know that the incessant fires, and the economic drain on their fiscal resources from constant firefighting, endanger the long-term survival of this small but important pollinator reserve.

Steve first visited the site and its environs near the little village of Bagaces in 1972 when it was known as "Palo Verde Station" and one of many destinations on the whirlwind educational tours organized by the Organization for Tropical Studies. The OTS program has trained thousands of U.S. and foreign biologists in tropical ecology through its "Fundamentals" and other courses. It is a consortium of member academic institutions based in San José, Costa Rica, and an exemplar of what can be done for teaching and research in the tropics. Almost all of the most highly respected biologists—even some lawyers and politicians—have slogged their way through its intensive dawn-to-dusk regimen. Before that, the place had been part of a large cattle and horse ranch, the Comelco property. During the late 1960s and early 1970s this area presented the ecotourist with unbroken vistas of gallery forest along unpolluted rivers and hillsides of huge canopy giants, strangler figs and epiphytic orchids, bromeliads, and others galore.

Today, most of those unbroken tracts have been taken over by various *vacas,* Big Macs, *caballos,* and charcoal. Only tiny isolated scraps and islandlike patches remain from what was only two or three decades ago a vast, forested, hilly, green, quilted landscape. Early faculty instructors and OTS students alike remember what it was like conducting pollination studies wavering on ladders or standing trembling on thin branches while wielding an insect net through the clouds of bees. In those days there were literally clouds of bees in the mind-blowing eye candy of mass-flowering legume trees, their combined wingbeats creating a ominous din that could be heard at some distance. The most spectacular of these trees burst forth in gaudy pink *(Tabebuia rosea)* or golden showers of intense lemon yellow *(Tabebuia ochracea)*. Amidst this sea of colorful tree crowns zoomed millions of small, medium, and large native bees, especially anthophorids, including the extremely fast genus *Centris.* In those days it was not difficult to collect long museum series or obtain specimens as pollination vouchers. Today, you often have to strain to hear the occasional bee visitor high overhead. The "Silent Spring" for Guanacaste bees has indeed occurred, not just for the singing of forest birds, but up in the nearly silent tree crowns now advertising for bees that no longer come and sip nectar.

The Lomas Barbudal Biological Reserve was the brainchild and far-sighted vision of a dedicated tropical biologist—Gordon W. Frankie of the Entomology Department at the University of California at Berkeley. Frankie has spent decades working in these forests studying when the plants bloom and which pollinators are attracted to reap their sweet rewards. Along with other colleagues he proposed that Lomas Barbudal could become a reserve to encompass 5,609 acres of dry forest lands and adjacent riparian gallery forest.

Why this site? Quite selfishly perhaps, they wanted to protect the research sites that had nurtured their careers and given their souls some respite from the concrete, steel, and glass world of large U.S. cities. They already knew that these forests were incredibly rich in native trees, shrubs, and herbs and that these flowering plants hosted an almost unimaginably rich guild of bird, bat, and insect pollinators. Bees were especially abundant here, and these biologists were enamored of the 20 or so giant anthophorid bees, fast-flying and beautiful members of the genus *Centris,* solitary nesters in both the ground and sometimes in the envelope surrounding termite nests. Most of these bees depend on the energy-rich floral oils provided by trees and lianas in the Malpighiaceae family. For bees this oil field is a bonanza. The dominant plant of these volcanic white hillsides, the "bee oil wells" as it were, is known by the name of "nance" *(Byrsonima crassifolia,)* and a wonderfully refreshing drink is prepared from its small yellow fruits. Its massive blossoms during the dry season, however, are what attract the centridine bees. Each flower bears calycine glands like miniature blisters of precious clear oil. Almost none of it goes to waste, either, as the bees squeegee it up and, in the dense coat of hairs on their hind legs, transport it back to their hidden nests. These floral oils—along with pollen from other plants—are literally turned into more *Centris* bees awaiting next year's mass blooming event.

Thus, this unique biological reserve was initially proposed by the Costa Rican biologists, various U.S. conservation agencies and numerous "Friends of Lomas Barbudal" as a bee and oil flower reserve that would ensure the long-term survival of the mother forest. We wish these courageous pioneers only the best in their efforts to curb clearing and other threats. Long may the *Centris* bees collect the floral oils for their larvae and, we hope, restore the din of buzzing wingbeats to the radiant Guanacaste countryside upon their return.

There are other large-scale efforts to establish or modify sites to serve as roosts for pollinators and provide them with forage nearby. In Arizona, for example, the U.S. Forest Service and National Park Service have devised a means of fencing off abandoned mine shafts in such a way that humans and wildlife cannot fall into them but nectar-feeding bats can continue to come and go on their night shifts. These mine shafts have now become essential to bat recovery plans, because many of the natural roosts for nectar-feeders have been disturbed by spelunkers, tourists, or vandals. Should such roost-site protection be coupled with nectar plant protection or reintroduction, bats will most certainly benefit from these actions.

In other regions of the United States, butterfly-attracting plants such as milkweeds are regularly mixed into roadside plantings. But as butterfly gardener Geyata Ajilvsgi has cautioned: "Unless highway right-of-ways are wide enough, all you end up with is mass killings—butterflies being hit by cars as they move out from small patches of wildflowers." Right-of-way plantings must either be wide, or congruous with native vegetation on the other side of the fence, so that butterflies find other attractive resources away from traffic.

To take Stanwyn Shetler's conservation message further, we must remember that a single patch of nectar plants does not a nectar corridor make. What we truly need in North America—and on other continents as well—are groups of skilled ecologists, gardeners, wildlife watchers, and restorationists dedicated to certain migratory corridors. They must record the typical period of migration for bat or monarch and list the floral resources locally available to the migrants during that period. They should ensure protection of these naturally available nectar sources and keep them free of herbicides and pesticides, especially when the migrants are soon to arrive.

More important, they need to link up with other groups of pollinator advocates farther down the line in order to spot any hazards or barriers along the nectar corridor. One exemplary environmental education project is now providing such a linkage, *on line,* to more than a thousand schools across North America. When Elizabeth Donnelly began to organize the Journey North program, she thought it would encourage schoolchildren "to observe phenological changes in their home place, to help them notice that there is not a 'silent spring' surrounding them." Now, several years later, Donnelly and her colleague Jennifer Gasperini at Hamline University have involved more than 25,000 children a day in tracking the journeys of 15 migratory species. The children are not only learning how state-of-the art technology can help them monitor the abundance and pace of migrants, they are also becoming advocates for these animals. As one teacher involved in the program put it: "My students suddenly saw the connection between the backyard butterfly, spring temperatures, and the distant country of Mexico. They were able to see how fragile nature is, and that all living things are interconnected."

Imagine—as participants in Chris Dresser's Migratory Species workshop did in April 1995—that schoolchildren could make a pilgrimage along such a nectar corridor, moving in synchrony with migrating pollinators. Imagine them inquiring of city fathers along the corridor why herbicides are still being used on milkweed patches in local highway medians, or why roosting trees in certain groves have not been protected from development. Imagine American children consulting with local botanists to learn which native milkweed to plant along each stretch of the corridor, or arriving in Michoacan to help Mexican schoolchildren plant oyamel firs or clean up local water holes essential to monarch overwintering roosts. If U.S. and Canadian children could see the "Fiesta Cultural de la Mariposa Monarca" in the small village of Angangueo—and learn of the rich oral history about monarchs remaining in Michoacan—perhaps it would inspire them to invite Mexican students to be their guests in the schools at the far northern end of the nectar corridor.

The same level of cultural commitment should be made, too, to nonmigratory pollinators and the native plants and crops that depend on them. A quarter century ago, zoologist Carl Johansen proposed that fieldside sanctuaries of native vegetation be established as "bee forage preserves" for certain native bees essential to crop yields in the northwestern United States. His proposal fell on deaf ears, however, for crop growers still expanding their fields into adjacent wildlands thought they could make up for the missing pollinators just by hiring migratory honeybee keepers rather than by protecting native bee habitat. Now that there are additional needs for maintaining wild pollinators, we hope that all groups can learn to work together to set aside areas for dwindling native bees and other insects such as butterflies and moths.

But now that the American honeybee industry has gone into a veritable tailspin, we are beginning to see that both the remaining honeybees and the native pollinators often require the wild forages adjacent to agricultural lands. We have had to learn, finally, from some hardwon wisdom in Western Europe. For many decades, European farmers have been trying to assess the ecological needs of native pollinators. As German wild bee conservationist Paul Westrich admits, they have long been faced with the problems of partial habitats. Today, he says,

> the entire habitat complex of a bee species very often consists of several partial habitats, [and] each partial habitat contains only one of the required resources. The nest site can lie several hundred meters away from the foraging site, and a completely different type of habitat not used by the bee can

divide them. . . . An intensive use of land very often leads to the destruction of one partial habitat and thereby to the loss of either the nesting or the foraging site. In consequence, the bee population . . . is extinguished. Therefore, protecting bees in many cases requires more than conserving just one site.

To conserve bees, Westrich and his German colleagues emphasize the importance of protecting cliffs, lowland heaths, marshes, deciduous woodlands, landslide-prone slopes, fens, riverbanks, sand dunes, and levees adjacent to farmlands. The European Community is finding that wildland protection provides more than recreation opportunities for humans; it is fundamental, as well, to sustaining agricultural productivity.

This message has finally taken root in American soil—or what is left of it—on Staten Island, New York. There, at the Fresh Kills Landfill, more than 3 billion cubic feet of garbage has been dumped over the last century, creating one of the largest human structures in North America. When the landfill is shut down in the year 2005, the mound of garbage there will have reached a height of 505 feet in elevation, the highest promontory on the Atlantic seaboard from Florida to Maine.

Over the last few years, Mary Yurlina and Steven Handel of Rutgers University have been involved in monitoring pollinator activity in a restoration experiment they hope will guide the ultimate revegetation of Fresh Kills. Small patches of native trees and shrubs have already been planted over the landfill, some of them more than a thousand feet away from the closest native woodland remaining on Staten Island. As Yurlina and Handel watched spring beauties and kinnikinnick dogwoods planted at various distances from the woodlands, they were surprised to find several species of bees visiting the flowers, even at the most distant plots. "These preliminary findings," they reported, "offer encouragement for a restoration of populations of important floral mutualists in newly established plant communities."

But when they monitored bee visitations to garden squash varieties at various distances from the remnant woodlands, the bee visitation rates declined on the squash blossoms furthest from the native vegetation. The message is clear: the fields and orchards that sustain our food supply should never become too far removed from wildlands, or their yields will suffer. The remaining wildlands and the animals that inhabit them are playing an increasingly important role in maintaining the stability of the world's food, fiber, and beverage supply. We cannot let wildness become too remote from the lives of pollinators, or from our own lives. The risk is too high—perhaps even higher than the mountain of garbage at the Fresh Kills Landfill.

Acosta-Ruiz, G., B. G. Carey, M.-A. Gonzalez, M. Gregory, B. Henson, and U. Navarro. 1994. *Border-Right-to-Know Project: The 1993 Northeastern Sonoran Pilot Inventories.* Bisbee: Arizona Toxics Information.

Adey, M., P. Walker, and P. T. Walker. 1986. *Pest Control Safe for Bees: A Manual and Directory for the Tropics and Subtropics.* London: International Bee Research Association.

Aizen, M. A. 1994. "Habitat fragmentation, native insect pollinators, and feral honey bees in Argentine 'Chaco Serrano.'" *Ecological Applications* 4(2):378–392.

Aizen, M. A., and P. Feinsinger. 1994. "Forest fragmentation, pollination, and plant reproduction in a Chaco Dry Forest, Argentina." *Ecology* 75(2):330–351.

Ajilvsgi, G. 1990. *Butterfly Gardening for the South.* Dallas: Taylor Publishing.

Alarcon, F. X. 1992. *Snake Poems.* San Francisco: Chronicle Books.

Allee, W. C. 1949. "Group survival value for *Philodina roseola,* a rotifer." *Ecology* 30:395–397.

Anderson, E. 1995. Personal communication to Gary Paul Nabhan regarding Galápagos prickly pear and others.

Anderson, J., and J. L. England. 1985. *Dwarf Bear-Poppy, Arctomecon humilis Coville Recovery Plan.* Denver: U.S. Fish and Wildlife Service.

Andres, T. C., and G. P. Nabhan. 1988. "Taxonomic rank and rarity of *Cucurbita okeechobeensis.*" *FAO/IBPGR Plant Genetic Resources Newsletter* 75/76:21–22.

Anonymous. 1994. "A binational partnership to protect Mexican free-tailed bats." *Bats* 12(4):6–7.

Arita, H. T., and C. Martinez del Rio. 1990. "Interacciones Flor-Murcielago: Un Enfoque Zoocentrico." *UNAM Instituto de Biología Publicaciones Especiales* 4:1–35.

Armstrong, J. A. 1979. "Biotic pollination mechanisms in the Australian flora: a review." *New Zealand Journal of Botany* 17:467–508.

Attenborough, D. 1995. *The Private Lives of Plants.* London: BBC Books.

Baker, H. G., and I. Baker. 1983. "Floral nectar sugar constituents in relation to pollinator type." In C. E. Jones and R. J. Little, eds., *Handbook of Experimental Pollination Biology.* Princeton: Van Nostrand-Reinhold.

———. 1990. "The predictive value of nectar chemistry to the recognition of pollinator types." *Israel Journal of Botany* 39:159–166.

Barth, F. G. 1985. *Insects and Flowers: The Biology of a Partnership.* Princeton: Princeton University Press.

Bawa, K. S. 1990. "Plant-pollinator interactions in tropical rainforests." *Annual Review of Ecology and Systematics* 21:399–422.

———. 1995. "Pollination, seed dispersal and diversification of angiosperms." *Trends in Ecology and Evolution* 10(8):311–312.

Bernhardt, P. 1987. "A comparison of the diversity, density, and foraging behavior of bees and wasps on Australian Acacia." *Annals of the Missouri Botanical Garden* 74:42–50.

———. 1989. *Wily Violets and Underground Orchids.* New York: Morrow.

Bernhardt, P., and L. B. Thien. 1987. "Self-isolation and insect pollination in the primitive angiosperms: new evaluations of older hypotheses." *Plant Systematics and Evolution* 156:159–176.

Bohart, G. E. 1972. "Management of wild bees for the pollination of crops." *Annual Review of Entomology* 17:287–312.

Bowers, J. E., and R. T. Turner. 1985. "A revised vascular flora of Tumamoc Hill, Tucson, AZ." *Madroño* 32:225–252.

Bowman, M.J.S. 1994. "Biodiversity: what does it mean?" *Biodiversity Letters* 2:1–3.

Boyd, R. S. 1994. "Pollination ecology of the rare shrub, Fremontodendron decumbens (*Sterculiaceae*)." *Madroño* 41:277–289.

Bronstein, J. L. 1994. Lecture notes, "Animal/Plant Interactions." Ecology and Evolutionary Department, University of Arizona, Tucson.

———. 1994. "The plant/pollinator landscape." In L. Hansson, L. Fahrig, and G. Merriam, eds., *Mosaic Landscapes and Ecological Processes.* London: Chapman & Hall.

Bronstein, J. L., P.-H. Gouyon, C. Gliddon, F. Kjellberg, and G. Michaloud. 1990. "The ecological consequences of flowering asynchrony in monoecious figs: a simulation study." *Ecology* 71(6):2145–2156.

Brower, L. P. 1993. "The influence of temperature on crawling, shivering, and flying in overwintering monarch butterflies in Mexico." In S. B. Malcolm and M. P. Zalucki, eds., *Biology and Conservation of the Monarch Butterfly.* Los Angeles: Los Angeles Natural History Museum.

Brown, D. E. 1994. *Vampire: The Vampire Bat in Fact and Fantasy.* Silver City, N.M.: High-Lonesome Books.

Bruman, H. 1949. "The cultural history of vanilla." *Hispanic-American Historical Review* 29:360–373.

Buchmann, S. L. 1996. "Specialist vs. generalist bees: efficiency as pollinators of Cucurbita foetidissima." Unpublished manuscript.

———. 1983. "Buzz pollination in angiosperms." In C. E. Jones and R. J. Little, eds., *Handbook of Experimental Pollination Biology.* Princeton: Van Nostrand-Reinhold.

———. 1987. "Floral biology of jojoba (*Simmondsia chinensis*), an anemophilous plant." *Desert Plants* 8(3):111–124.

———. 1987. "The ecology of oil flowers and their bees." *Annual Review of Ecology and Systematics* 18:343–369.

———. 1994. "Beekeeping without honey bees: attracting other bees to your garden." *Invertebrates in Captivity Conference Proceedings.* Tucson: SASI.

Buchmann, S. L., and M. D. Buchmann. 1981. "Anthecology of *Mouriri myrtilloides* (Melastomataceae: Memecyleae), an oil flower in Panama." *Biotropica* 13(2):7–24.

Buchmann, S. L., and J. H. Cane. 1989. "Bees assess pollen returns while sonicating Solanum flowers." *Oecologia* 81:289–294.

Buchmann, S. L., and M. K. O'Rourke. 1988. "Palynological analysis of dietary breadth of honey bee colonies in the Sonoran Desert (1980–1987)." *Proceedings of the XVIII International Congress of Entomology.* Vancouver, B.C.: ICE.

———. 1991. "Importance of pollen grain volumes for calculating bee diets." *Grana* 30:591–595.

Buchmann, S. L., and C. W. Shipman. 1990. "Pollen harvesting rates for *Apis mellifera* L. on *Gossypium* (Malvaceae) flowers." *Journal of the Kansas Entomological Society* 63(1):92–100.

Buchmann, S. L., M. K. O'Rourke, C. W. Shipman, S. C. Thoenes, and J. O. Schmidt. 1993. "Pollen harvest by honey bees in Saguaro National Monument: potential effects on plant reproduction." In C. P. Stone and E. S. Bellantoni, eds., *Proceedings of the Symposium on Research in Saguaro National Monument.* Tucson: National Park Service.

Burd, M. 1994. "Bateman's principle and plant reproduction: the role of pollen limitation in fruit and seed set." *Botanical Review* 60:81–109.

Burger, W. C. 1981. "Why are there so many flowering plants?" *BioScience* 31(8):572–577.

Burgman, M. A., S. Ferson, and H. R. Akcakaya. 1992. *Risk Assessment in Conservation Biology.* London: Chapman & Hall.

Cane, J. H. 1994. "Nesting biology and mating behavior of the southeastern blueberry bee, *Habropoda laboriosa* (Hymenoptera: Apoidea)." *Journal of the Kansas Entomological Society* 67(3):236–241.

Cane, J. H., R. R. Snelling, L. J. Kervin, and G. C. Eickwort. 1996. "A new monolectic coastal bee, *Hesperapis oraria* Snelling & Stage (Hymenoptera: Melittidae), with a review of desert disjunctives in the southeastern U. S." *Journal of the Kansas Entomological Society* (in press).

Carlquist, S. 1974. *Island Biology*. New York: Columbia University Press.

Carson, R. 1962. *Silent Spring*. New York: Houghton Mifflin.

Ceballos, G., and J. H. Brown. 1995. "Global patterns of mammalian diversity, endemism and endangerment." *Conservation Biology* 9(3):559–568.

Center for Plant Conservation. 1992. *List of Plant Species of Concern in the United States*. St. Louis: Missouri Botanical Garden.

Collar, N. J., M. J. Crosby and A. J. Stattersfield. 1994. *Birds to Watch 2: The World List of Threatened Birds*. Washington, D.C.: Birdlife International, Smithsonian Institution.

Cory, C. 1984. "Pollination biology of two species of Hawaiian Lobeliaceae *(Clermontia kakeana* and *Cyanea angustifolia)* and their presumed co-evolved relationship with native honeycreepers *(Drepanididae)*." Master's thesis, University of California–Davis.

Cox, P. A., T. Elmqvist, E. D. Pierson, and W. E. Rainey. 1991. "Flying foxes as strong interactors in South Pacific Island ecosystems: a conservation hypothesis." *Conservation Biology* 5(4):448–454.

Crane, E. 1983. *The Archaeology of Beekeeping*. Ithaca: Cornell University Press.

Curtis, J., T. Profeta, and L. Mott. 1993. *After Silent Spring*. Washington, D.C.: Natural Resources Defense Council.

Davidson, M. 1988. *Convictions of the Heart*. Tucson: University of Arizona Press.

Dinham, B. (ed.). 1993. *The Pesticide Hazard*. Atlantic Highlands, N.J.: Zed Books.

Dodson, C. H., R. L. Dressler, H. G. Hills, R. M. Adams, and N. H. Williams. 1959. "Biologically active compounds in orchid fragrances." *Science* 164:1243–1249.

Doyle, J. A. 1977. "Patterns of evolution in early angiosperms." In A. Hallam, ed., *Patterns of Evolution*. Amsterdam: Elsevier.

Eguiarte, L. E., and A. Burquez. 1988. "Reducción en la fecundidad en *Manfreda brachystachya* (Cav.) Rose, un agavacaea polinizada por murcielagos: los riesgos de la especializacion en la polinizacion." *Boletín de la Sociedad Botanica Mexicana* 48:147–149.

Ehrlich, P. R., D. S. Dobkin, and D. Wheye. 1992. *Birds in Jeopardy.* Stanford: Stanford University Press.

Eickwort, G. C., and H. S. Ginsberg. 1980. "Foraging and mating behavior in Apoidea." *Annual Review of Entomology* 25:421–446.

Ellstrand, N. C., and C. A. Hoffman. 1990. "Hybridization as an avenue of escape for engineered genes." *BioScience* 40(6):438–442.

Elmqvist, T., P. A. Cox, W. E. Rainey, and E. D. Pierson. 1992. "Restricted pollination on oceanic islands: pollination of *Ceiba acuminata* by flying foxes in Samoa." *Biotropica* 24(1):15–23.

Falk, D., and K. Holsinger. 1992. *Genetics and Conservation of Rare Plants.* Oxford: Oxford University Press.

Feinsinger, P. 1983. "Coevolution and pollination." In D. J. Futuyma and M. Slatkin, eds., *Coevolution.* Sutherland, Mass.: Sinauer.

———. 1990. "Interacciones entre plantas y colibries en selvas tropicales." *Boletín de la Academia Nacional de Ciencias* 59(1–2):31–54.

Feinsinger, P., J. A. Wolfe, and L. A. Swarm. 1982. "Island ecology: reduced hummingbird diversity and the pollination biology of plants, Trinidad and Tobago, West Indies." *Ecology* 63(2):494–506.

Feinsinger, P., J. H. Beach, Y. Linhart, W. H. Rusby, and K. G. Murray. 1987. "Disturbance, pollinator predictability, and pollination success among Costa Rican cloud forest plants." *Ecology* 68(5):1294–1305.

Fiedler, P. 1993. "Habitat fragmentation and its demographic consequences: overview and consequences." In O. T. Sandlund and P. J. Schel, eds., *Proceedings of the Norway/UNEP Expert Conference on Biodiversity.* Trondheim: Norwegian Ministry of the Environment.

Fleming, T. H. 1993. "Plant-visiting bats." *American Scientist* 81:461–468.

Fleming, T. H., R. A. Nuñez, and L. da Silviera Lobo Sternberg. 1993. "Seasonal changes in the diets of migrant and non-migrant nectarivorous bats as revealed by carbon stable isotope analysis." *Oecologia* 94:72–75.

Ford, H. A., D. C. Paton, and N. Forde. 1979. "Birds as pollinators of Australian plants." *New Zealand Journal of Botany* 17:509–519.

Free, J. B. 1993. *Insect Pollination of Crops.* 2nd ed. London: Academic Press.

Fritz, A.-L., and L. A. Nilsson. 1994. "How pollinator-mediated mating varies with population size in plants." *Oecologia* 100:451–462.

Ginsberg, H. S. 1983. "Foraging ecology of bees in an old field." *Ecology* 54:165–175.

Grant, V. 1994. "Historical development of ornithophily in the western North American flora." *Proceedings of the National Academy of Sciences, USA* 91:10407–10411.

Grant, V., and K. A. Grant. 1983. "Behavior of hawkmoths on flowers of *Datura metaloides.*" Botanical Gazette 144:280–284.

Griffin, B. L. 1993. *The Orchard Mason Bee*. Bellingham, Wash.: Knox Cellars Publishing.

Griswold, T. 1993. "New species of *Perdita (Pygoperdita)* Timberlake of the *P. californica* species group (Hymenoptera: Andrenidae)." *Pan-Pacific Entomologist* 69:183–189.

Hafernik, Jr., J. E. 1992. "Threats to invertebrate biodiversity: implications for conservation strategies." In P. L. Feidler and S. K. Jain, eds., *Conservation Biology*. London: Chapman & Hall.

Hagler, J. R., and S. L. Buchmann. 1993. "Honey bee (Hymenoptera: Apidae) foraging responses to phenolic-rich nectars." *Journal of the Kansas Entomological Society* 66(2):223–230.

Hanscom, T., and W. Toone. 1995. "Economic independence and conservation in the tropics." *Wings* 18:3–5.

Heinrich, B. 1975. "Energetics of pollination." *Annual Review of Ecology and Systematics* 6:139–170.

———. 1979. *Bumblebee Economics*. Cambridge: Harvard University Press.

Hendrix, S. D. 1994. "Effects of population size on fertilization, seed production, and seed predation in two prairie legumes." *North American Prairie Conference Proceedings* 13:115–119.

Hendrix, S. D., R. W. Cruden, and B. M. Molano-Flores. 1993. *Effects of Landscape Fragmentation on the Reproductive Biology of* Phlox pilosa. Ecological Society of America Abstract 273. Washington, D.C.: ESA.

Hennesssey, M. K., H. N. Higg, and D. H. Habeck. 1992. "Mosquito (Diptera: Culicidae) adulticide spray drift into wildlife refuges of the Florida Keys." *Environmental Entomology* 21:714–721.

Hilty, S. 1994. *Birds of Tropical America*. Shelburne, Vt.: Chapters.

Howell, D. J. 1974. "Pollinating bats and plant communities." *National Geographic Society Research Reports* 1373:311–328.

Hughes, N. F. 1976. *Paleobiology of Angiosperm Origins*. New York: Cambridge University Press.

Hurd, Jr., P. D., and E. G. Linsley. 1964. "The squash and gourd bees—genera *Peponapis* Robertson and *Xenoglossa* Smith—inhabiting America north of Mexico (Hymenoptera: Apoidea)." *Hilgardia* 35(15):375–476.

Hurd, Jr., P. D., E. G. Linsley, and T. W. Whitaker. 1971. "Squash and gourd bees *(Peponapsis, Xenoglossa)* and the origin of cultivated *Cucurbita*." *Evolution* 25:218–234.

Janson, C. H., J. Terborgh, and L. H. Emmons. 1981. "Non-flying mammals as pollinating agents in the Amazonian forest." *Biotropica* 13:1–6.

Janzen, D. H. 1974. "The de-flowering of Central America." *Natural History* 83:49.

Jennersten, O. 1988. "Pollination of *Dianthus deltoides* (Caryophyllaceae): effects of habitat fragmentation on visitation and seed set." *Conservation Biology* 2:359–366.

Jennersten, O., J. Loman, A. P. Muller, J. Robertson, and B. Widen. 1992. "Conservation biology in agricultural habitats." In J. Harrison, ed., *Ecological Perspectives of Nature Conservation*. London: Elsevier.

Johansen, C. A. 1969. "Bee forage preserves." *American Bee Journal* 109:96–97.

―――. 1977. "Pesticides and pollination." *Annual Review of Entomology* 22:177–192.

Johansen, C. A., and D. F. Mayer. 1990. *Pollinator Protection: A Bee and Pesticide Handbook*. Cheshire, Conn.: Wicwas Press.

Kevan, P. G. 1974. "Bees, blueberries, birds and budworm." *Osprey: Newfoundland Natural History Society Newsletter* 5(3):54–62.

―――. 1975. "Pollination and environmental conservation." *Environmental Conservation* 2(4):293–298.

―――. 1977. "Blueberry crops in Nova Scotia and New Brunswick—pesticides and crop reductions." *Canadian Journal of Agricultural Economics* 25(1):64.

―――. 1994. *Ecologica Evolutiva de la Polinizacíon*. Mexico City: UNAM Instituto de Biología.

―――. 1994. "Honey hunting and gathering: a Malaysian tropical forest expedition for the adventurous tourist." *American Bee Journal* 134:41–42.

Kevan, P. G., E. A. Clark, and V. G. Thomas. 1990. "Insect pollinators and sustainable agriculture." *American Journal of Alternative Agriculture* 5(1):13–22.

Klinger, T., D. R. Elam, and N. C. Ellstrand. 1991. "Radish as a model system for the study of engineered gene escape rates via crop-weed mating." *Conservation Biology* 5(4):531–535.

Kohn, J. R. 1988. "Why be female?" *Nature* 335:431–433.

Koopowitz, H. 1992. "A stochastic model for the extinction of tropical orchids." *Selbyana* 13:115–122.

Koopowitz, H., A. Thornhill, and M. Andersen. 1993. "Species distribution profiles of the neotropical orchids *Masdevallia* and *Dracula* (Pleurothallidinae, Orchidaceae): implications for conservation." *Biodiversity and Conservation* 2:681–690.

―――. 1994. "A general stochastic model for the prediction of biodiversity losses based on habitat conversion." *Conservation Biology* 8(2):425–438.

Koopowitz, H., M. Andersen, A. Thornhill, H. Nguyen, and A. Pham. 1993. "Comparison of distributions of terrestrial and epiphytic African orchids: implications for conservation." In A. Pridgeon, ed., *Proceedings of the Fourteenth World Orchid Conference*. Edinburgh: HMSO.

Kress, J. W., and J. B. Beach. 1994. "Flowering plant reproductive systems." In L. A. McDade, K. S. Bawa, H. A. Hespenheide, and G. Hartshorn, eds., *La*

*Selva: Ecology and Natural History of a Neotropical Rain Forest.* Chicago: University of Chicago Press.

Kress, J. W., G. E. Schatz, M. Andrianifahanana, and H. S. Morland. 1994. "Pollination of *Ravenala madagascarensis* (Sterlitziaceae) by lemurs in Madagascar: evidence for an archaic coevolutionary system?" *American Journal of Botany* 81(5):542–551.

LaMont, B. B., P.G.L. Klinkhamer, and E.T.F. Witkowski. 1993. "Population fragmentation may reduce fertility to zero in *Banksia goodii*—a demonstration of the Allee effect." *Oecologia* 94:446–450.

LaSalle, J., and I. D. Gauld (eds.). 1993. *Hymenoptera and Biodiversity.* Wallingford, England: CAB International.

Lesica, P. 1993. "Loss of fitness resulting from pollinator exclusion in *Silene spaldingii* (Caryophyllaceae)." *Madroño* 40(4):193–201.

Linhart, Y. B., and P. Feinsinger. 1980. "Plant–hummingbird interactions: effects of island size and degree of specialization on pollination." *Journal of Ecology* 68:745–760.

Linsley, E. G. 1966. "Pollinating insects of the Galapagos Islands." In R. I. Bowman, ed., *The Galapagos.* Berkeley: University of California Press.

Lord, J. M. 1991. "Pollination and seed dispersal in *Freycinetia baueriana,* a dioecious liana that has lost its pollinator." *New Zealand Journal of Botany* 29:83–86.

Lugo, A. 1988. "Estimating reductions in the diversity of tropical forest species." In E. O. Wilson, ed., *Biodiversity.* Washington, D.C.: National Academy Press.

MacGregor, S. E. 1976. *Insect Pollination of Cultivated Crop Plants.* USDA Agriculture Handbook 496. Washington, D.C.: USDA.

Matheson, A., S. L. Buchmann, C. O' Toole, P. Westrich, and I. H. Williams (eds.) 1996. *The Conservation of Bees.* Academic Press, London: In press.

May, R. M., and A. M. Lyles. 1987. "Living Latin binomials." *Nature* 326:642–644.

McKey, D. 1989. "Population biology of figs: applications for conservation." *Experientia* 45:661–673.

Medellin-Morales, S., and M. M. Cruz-Bojorquez. 1992. *Xunan Kab: Una experencia etnoecologica y de transferencia de tecnologia tradicionál en una comunidad Maya de Yucatan.* Proyecto. Sostenibilidad Maya, University of California–Riverside.

Medellin-Morales, S., E. Campos-Lopez, E. Campos-Nanez, J. Gonzalez-Acereto, and V. Camara-Gonzalez. 1991. *Meliponicultura Maya: Perspectivas Para Su Sostenibilidad.* Proyecto Sostenibilidad Maya, University of California–Riverside.

Meeuse, B., and S. Morris. 1984. *The Sex Life of Flowers*. New York: Facts on File.

Menges, E. S. 1991. "Seed germination percentage increases with population size in a fragmented prairie species." *Conservation Biology* 5:158–164.

Merrick, L. C., and G. P. Nabhan. 1985. "Natural hybridization of wild *Cucurbita sororia* group and domesticated *Cucurbita mixta* in southern Sonora." *Cucurbit Genetic Cooperative Newsletter* 7:73–75.

Michener, C. D. 1974. *The Social Behavior of Bees*. Cambridge: Belknap Press/Harvard University Press.

Michener, C. D., R. J. McGinley, and B. N. Danforth. 1994. *The Bee Genera of North and Central America (Hymenoptera: Apoidea)*. Washington, D.C.: Smithsonian Institution Press.

Mills, S. 1995. *In Service of the Wild: Restoring and Reinhabiting Damaged Land*. Boston: Beacon Press.

Mittermeier, R. A., I. Tattersall, W. R. Konstant, D. M. Meyers, and R. B. Mast. 1994. *Lemur of Madagascar*. Washington, D.C.: Conservation International.

Nabhan, G. P. 1973. "A characterization of Galápagos mangrove communities in relation to coastal geomorphological evolution." Unpublished honors thesis, Prescott College, Arizona.

———. 1984. "Evidence of gene flow between cultivated *Cucurbita mixta* and a field edge population of wild *Cucurbita* in Onavas, Sonora." *Cucurbit Genetics Cooperative Newsletter* 7:76–77.

———. 1989. *Enduring Seeds: Native American Agriculture and Wild Plant Conservation*. San Francisco: North Point Press.

———. 1995. "The dangers of reductionism in biodiversity conservation." *Conservation Biology* 9(3):479–481.

Nabhan, G. P., and S. Buchmann. 1995. "Chemically-induced habitat fragmentation and disrupted plant/pollinator relationships." *Comments in Toxicology*. In press.

Nabhan, G. P., and T. Fleming. 1993. "The conservation of mutualisms." *Conservation Biology* 7(3):457–459.

Nabhan, G. P., and H. Suzan. 1994. "Boundary effects on endangered cacti and their nurse plants in and near a Sonoran Desert biosphere reserve." In G. P. Nabhan and J. L. Carr, eds., *Ironwood: An Ecological and Cultural Keystone of the Sonoran Desert*. Conservation International Occasional Papers in Conservation Biology 1. Washington, D.C., Conservation International.

National Association of State Departments of Agriculture. 1991. *Honey Bee Pests — A Threat to the Vitality of U.S. Agriculture: A National Strategy*. Washington, D.C.: NASDA.

Neff, J. L., and B. B. Simpson. 1993. "Bees, pollination systems and plant diversity." In J. LaSalle and I. D. Gauld, eds., *Hymenoptera and Biodiversity.* Wallingford, England: CAB International.

New, T. R., R. M. Pyle, J. A. Thomas, and P. Hammond. 1995. "Butterfly conservation management." *Annual Review of Entomology* 40:57–83.

Niklas, K. J. 1992. *Plant Biomechanics: An Engineering Approach to Plant Form and Function.* Chicago: University of Chicago Press.

Niklas, K. J., and S. L. Buchmann. 1985. "Aerodynamics of wind pollination in *Simmondsia chinensis.*" *American Journal of Botany* 72(4):530–539.

————. 1987. "The aerodynamics of pollen capture in two sympatric *Ephedra* species." *Evolution* 41(1):104–123.

Nilsson, L. A. 1992. "Long pollinia on eyes: hawk-moth pollination of *Cynorkis uniflora* Lindley (Orchidaceae) in Madagascar." *Botanical Journal of the Linnean Society* 109:145–160.

Noss, R. F., and A. Y. Cooperrider. 1994. *Saving Nature's Legacy.* Washington, D.C.: Island Press.

Olesen, J. M., and S. K. Jain. 1994. "Fragmented populations and their lost interactions." In V. Lennische, J. Tannah, and S. K. Jain, eds., *Conservation Genetics.* Bwerkes-Verlag, Switzerland.

O'Neal, R. J., and G. D. Waller. 1982. "On the pollen harvest by the honey bee *(Apis mellifera L.)* near Tucson, Arizona (1976–1981)." *Desert Plants* 6:81–109.

Ordway, E., S. L. Buchmann, R. O. Kuehl, and C. W. Shipman. 1987. "Pollen dispersal in *Curcubita foetidissima* (Cucurbitaceae) by bees of the genera *Apis, Peponapis* and *Xenoglossa* (Hymenoptera: Apidae, Anthophoridae)." *Journal of the Kansas Entomological Society* 60(4):489–503.

Ornelas, J. F. 1994. "Serrate tomia: an adaptation for nectar-robbing in hummingbirds?" *The Auk* 11(3):703–710.

O'Rourke, M. K., and S. L. Buchmann. 1991. "Standardized analytical techniques for bee-collected pollen." *Environmental Entomology* 20(2):507–513.

O'Toole, C. 1993. "Diversity of native bees and agroecosystems." In J. LaSalle and I. D. Gauld, eds., *Hymenoptera and Biodiversity.* Wallingford, England: CAB International.

O'Toole, C., and A. Raw. 1991. *Bees of the World.* New York: Facts on File.

Parker, F. D., and P. F. Torchio. 1980. "Management of wild bees." In USDA Agriculture Handbook 335. Washington, D.C.: USDA.

Parker, F. D., S.W.T. Batra, and V. J. Tepedino. 1987. "New pollinators for our crops." *Agricultural Zoology Reviews* 2:279–304.

Paton, D. C. 1990. "Budgets for the use of floral resources in mallee-heath." In J. C. Noble, P. J. Joss, and G. K. Jones, eds., *The Mallee Lands: A Conservation Perspective.* Melbourne: CSIRO.

————. 1993. "Honeybees in the Australian environment." *BioScience* 43(2):95–103.

————. 1995. "Overview of feral and managed honeybees in Australia: distribution, abundance, extent of interactions with native biota, evidence of impacts and future research." University of Adelaide Zoology Department, Adelaide, Australia.

Pavlik, B. M., N. Ferguson, and M. Nelson. 1993. "Assessing limitations on the growth of endangered plant populations, II: Seed production and seed bank dynamics of *Erysimum capitatum* ssp. *angustatum* and *Oenothera deltoides* ssp. *howellii.*" *Biological Conservation* 65:267–278.

Pavlik, B. M., N. Ferguson, E. Manning, and M. G. Barbour. 1988. *Demographic Studies of Endemic Plants at the Antioch Dunes National Wildlife Refuge.* Vol. I: *Seed Production and Germination.* Sacramento: California Department of Fish and Game.

Pellmyr, O. 1989. "The cost of mutualism: interactions between *Trollius europaeus* and its pollinating parasites." *Oecologia* 78:53–59.

Pellmyr, O., and C. J. Huth. 1994. "Evolutionary stability of mutualism between yuccas and yucca moths." *Nature* 372:257–260.

Petanidou, T., and W. N. Ellis. 1993. "Pollinating fauna of a phryganic ecosystem: composition and diversity." *Biodiversity Letters* 1:9–22.

Phillips, O. L., and A. H. Gentry. 1994. "Increasing turnover through time in tropical forests." *Science* 263:954–958.

Phillips, O. L., and B. Meilleur. 1995. *Survey of the Usefulness and Economic Potential of the Rare Plants of the United States.* Final Report. St. Louis: Center for Plant Conservation.

Primack, R. B., and P. Hall. 1990. "Costs of reproduction in the pink lady's slipper orchid: a four-year experimental study." *American Naturalist* 136(5):638–656.

Prys-Jones, O. E., and S. A. Corbet. 1991. "Bumblebees." In *Naturalists' Handbook No. 6.* Slough, England: Richmond Publishing.

Przeslawski, J. 1995. "The end of the never-ending forest." *American Orchid Society Bulletin* (January):44–49.

Pyke, G. H. 1983. "Relationships between honeyeater numbers and nectar production in heathlands near Sydney." *Australian Journal of Ecology* 5:343–370.

————. 1990. "Apiarists versus scientists: a bittersweet case." *Australian Natural History* 23:386–392.

Pyke, G. H., and L. Balzer. 1985. *The Effect of the Introduced Honey-bee on Australian Native Bees.* Occasional Paper 7. Sydney: New South Wales National Parks and Wildlife Service.

Pyke, G. H., and H. F. Recher. 1986. "Relationship between nectar production and seasonal patterns of density and nesting of resident honeyeaters in heathland near Sydney." *Australian Journal of Ecology* 11:195–200.

Pyke, G. H., and N. M. Waser. 1981. "The production of dilute nectars by hummingbird and honeyeater flowers." *Biotropica* 13:65–66.

Pyle, R. 1993. *The Thunder Tree*. New York: Houghton Mifflin.

Quammen, D. 1996. *Song of the Dodo*. New York: Knopf.

Rappole, J. H., E. S. Morton, T. E. Lovejoy III, and J. L. Ruos. 1993. *Aves Migratorias Nearticas en Los Neotropicos*. Washington, D.C.: Smithsonian Institution.

Rathcke, B. J., and E. Jules. 1993. "Habitat fragmentation and plant/pollinator interactions." *Current Science* 65:273–278.

Rathje, W., and C. Murphy. 1992. *Rubbish! The Archaeology of Garbage*. New York: HarperPerennial.

Regal, P. J. 1977. "Ecology and evolution of flowering plant dominance." *Science* 196:622–629.

Reidinger, Jr., R. 1976. "Organochlorine residues in adults of six southwestern bat species." *Journal of Wildlife Management* 40(4):677–680.

Rick, C. M. 1966. "Some plant/animal relations on the Galapagos Islands." In R. I. Bowman, ed., *The Galapagos*. Berkeley: University of California Press.

Robinson, W. S., R. Nowogrodski, and R. A. Morse. 1989. "The value of honey bees as pollinators of U.S. crops." Parts I and II. *American Bee Journal* 129:411–423, 477–487.

Rondeau, R., and T. R. VanDevender, P. D. Jenkins, C. D. Bertelsen, and R. K. VanDevender. 1993. "Flora of the Tucson Mountains." In *Proceedings of the Symposium on Research in Saguaro National Monument*. Tucson: National Park Service.

Rothschild, M. 1985. "The flowering hayfield." *House and Garden* (January):24–25.

Rothschild, M., and C. Farrell. 1983. *The Butterfly Gardener*. London: Rainbird.

Roubik, D. W. 1978. "Competitive interactions between neo-tropical pollinators and Africanized honey bees." *Science* 201:1030–1032.

———. 1980. "Foraging behavior of competing Africanized honeybees and stingless bees." *Ecology* 63:836–845.

———. 1982. "Ecological impact of Africanized honey bees on native neotropical pollinators." In P. Jaisson, ed., *Social Insects in the Tropics*. Paris: Presses de l'Université Paris XIII.

———. 1983. "Experimental community studies: time series tests of competition between African and neotropical bees." *Ecology* 64:971–978.

————. 1989. *Ecology and Natural History of Tropical Bees.* Cambridge: Cambridge University Press.

————. 1990. "Niche preemption in tropical bee communities: a comparison of neotropical and malesian faunas." In S. F. Sakagami, R. Ohgushi, and D. W. Roubik, eds., *Natural History of Social Wasps and Bees in Equatorial Sumatra.* Sapporo: Hokkaido University Press.

————. 1991. "Aspects of Africanized honey bee ecology in tropical America." In M. Spivak, D.J.C. Fletcher, and M. D. Breed, eds., *The African Honey Bee.* Boulder: Westview Press.

————. 1995. *Pollination of Cultivated Plants in the Tropics.* Agricultural Services Bulletin 118. Rome: Food and Agriculture Organization of the United Nations.

Roubik, D. W., and J. D. Ackerman. 1987. "Long-term ecology of euglossine orchid-bees (Apidae: Euglossini) in Panama." *Oecologia* 73:321–33.

Roubik, D. W., and S. L. Buchmann. 1984. "Nectar selection by *Melipona* and *Apis mellifera* (Hymenoptera: Apidae) and the ecology of nectar intake by bee colonies in a tropical forest." *Oecologia* 61:1–10.

Roubik, D. W., J. E. Moreno, C. Vergara, and D. Wittmann. 1986. "Sporadic food competition with the African honey bee: projected impact on neotropical social bees." *Journal of Tropical Ecology* 2:97–111.

Schaffer, W. M., D. B. Jensen, D. E. Hobbs, J. Gurevitch, J. R. Todd, and M. V. Schaffer. 1979. "Competition, foraging energetics, and the cost of sociality in three species of bees." *Ecology* 60(5):976–987.

Schaffer, W. M., D. W. Zeh, S. L. Buchmann, S. Kleinhans, M. V. Schaffer, and J. Antrim. 1983. "Competition for nectar between introduced honey bees and native North American bees and ants." *Ecology* 64(3):564–577.

Schmalzel, R. 1980. "The Diet Breadth of *Apis* (Hymenoptera: Apidae)." Master's thesis, University of Arizona, Tucson.

Schmidt, J. O., and S. L. Buchmann. 1986. "Floral biology of the saguaro *(Cereus giganteus).* I: Pollen harvest by *Apis mellifera.*" *Oecologia* 69:491–498.

Schrader, E. 1995. "A giant spraying sound." *Mother Jones* (January/February):34–36, 73–74.

Seeley, T. D. 1985. *Honeybee Ecology: A Study of Adaptation in Social Life.* Princeton: Princeton University Press.

Sipes, S. D., and V. J. Tepedino. 1995. "Reproductive biology of the rare orchid, *Spiranthes diluvialis:* breeding system, pollination, and implications for conservation." *Conservation Biology* 9(4):929–938.

Sipes, S. D., V. J. Tepedino, and W. R. Bowlin. 1993. "The pollination and reproductive ecology of *Spiranthes diluvialis* Sheviak (Orchidaceae)." In

R. Sivinski and K. Lightfoot, eds., *Proceedings of the Southwestern Rare and Endangered Plant Conference.* Santa Fe: New Mexico Forestry Dept.

Sisk, T. D., A. E. Launer, K. R. Switky, and P. R. Ehrlich. 1994. "Identifying extinction threats." *BioScience* 44(9):592–604.

Sladen, F.W.L. 1989. *The Humble-Bee: Its Life History and How to Domesticate It.* Worcester, England: Logaston Press.

Smith, T. B., L. A. Freed, J. K. Lepson, and J. H. Carothers. 1995. "Evolutionary consequences of extinctions in populations of a Hawaiian honeycreeper." *Conservation Biology* 9:107–113.

Soberon-Mainero, J., and C. Martinez del Rio. 1985. "Cheating and taking advantage of mutualistic associations." In D. Boucher, ed., *The Biology of Mutualism.* Oxford: Oxford University Press.

Southwick, E. E., and L. Southwick, Jr. 1989. "A comment on 'Value of honey bees as pollinators of U.S. crops.'" *American Bee Journal* 129:805–807.

———. 1992. "Economic value of honey bees (Hymenoptera: Apidae) in the United States." *Journal of Economic Entomology* 85(3):621–633.

Spears, E. E. 1987. "Island and mainland pollination ecology of *Centrosema virginianum* and *Opuntia stricta.*" *Journal of Ecology* 75:351–362.

Steiner, J. 1993. "Lange's remarkable metalmark." *Tideline* 13(1):1–4.

Stephen, W. P., G. E. Bohart, and P. F. Torchio. 1969. *The Biology and External Morphology of Bees: With a Synopsis of the Genera of Northwestern America.* Corvallis: Oregon State University, Agricultural Experiment Station.

Stolzenburg, W. 1992. "The lonesome flower." *Nature Conservancy News* (March/April):28–30.

Sugden, E. A., and G. H. Pyke. 1991. "Effects of honeybees on colonies of *Exoneura asimillima,* an Australian native bee." *Australian Journal of Ecology* 16:171–181.

Sugden, E. A., R. W. Thorp, and S. L. Buchmann. 1995. "Honey bee–native bee competition in Australia: focal point for environmental change and apicultural response." *Bee World* (in press).

Suzan, H., G. P. Nabhan, and D. T. Patten. 1994. "Nurse plant and floral biology of a rare night-blooming cereus, *Peniocereus striatus* (Brandegee) Buxbaum." *Conservation Biology* 8:461–470.

Tekulsky, M. 1985. *The Butterfly Garden.* Boston: Harvard Common Press.

———. 1990. *The Hummingbird Garden.* New York: Crown Publishers.

Temple, S. A. 1977. "Plant-animal mutualism: coevolution with dodo leads to near extinction of plant." *Science* 197:885–886.

Tepedino, V. J. 1979. "The importance of bees and other insect pollinators in maintaining floral species composition." *Great Basin Naturalist Memoirs* 3:139–150.

————. 1981. "The pollination efficiency of the squash bee *(Peponapis pruinosa)* and the honey bee *(Apis mellifera)* on summer squash *(Cucurbita pepo)." Journal of the Kansas Entomological Society* 54(2):359–377.

Thien, L. B. 1980. "Patterns of pollination in the primitive angiosperms." *Biotropica* 12(1):1–13.

Thomas, J. A. 1991. "Rare species conservation: case studies of European butterflies." *Symposium of the British Ecological Society* 31:149–197.

Thomas, K. 1984. *Man and the Natural World.* Oxford: Oxford University Press.

Tilman, D., R. M. May, C. L. Lehman, and M. A. Novak. 1994. "Habitat destruction and the extinction debt." *Nature* 371:65–66.

Toledo, V. M. 1977. "Pollination of some rain forest plants by non-hovering birds in Veracruz, Mexico." *Biotropica* 9(4):262–267.

Torchio, P. F. 1973. "Relative toxicity of insecticides to the honey bee, alkali bee, and alfalfa leafcutting bee." *Journal of the Kansas Entomological Society* 46(4):446–453.

Torchio, P. F., T. L. Griswold, and F. Messina. 1994. "Biology and floral fidelity of *Perdita meconis,* a pollinator of a rare plant." *USDA Abstracts* 9146965. Washington, D.C.: USDA.

Tuttle, M. 1990. *America's Neighborhood Bats.* Austin: University of Texas Press.

Vane-Wright, R. I. 1993. "The Columbus hypothesis: an explanation for the dramatic 19th century expansion of the monarch butterfly." In S. B. Malcolm and M. P. Zalucki, eds., *Biology and Conservation of the Monarch Butterfly.* Los Angeles: Los Angeles Natural History Museum.

Verey, R. 1985. "The wild bunch." *House and Garden* (January):188–189.

Vinson, S. B., G. W. Frankie, and J. Barthell. 1994. "Threats to the diversity of solitary bees in a neotropical dry forest in Central America." In J. LaSalle and I. D. Gauld, eds., *Hymenoptera and Biodiversity.* Wallingford, England: CAB International.

Walters, T. W., and D. Decker-Walters. 1993. "Systematics of the endangered Okeechobee gourd, *(Cucurbita okeechobeensis,* Cucurbitaceae)." *Systematic Botany* 18(2):175–187.

Waser, N. M. 1979. "Pollinator availability as a determinant of flowering time in ocotillo *(Foquieria splendens)." Oecologia* 39:107–121.

————. 1984. "Pollen shortcomings." *Natural History* 93:26–31.

Waser, N. M., and L. A. Real. 1979. "Effective mutualism between sequentially flowering plant species." *Nature* 281:670–672.

Waser, N. M., M. Price, and R. Menzel. 1996. "Specialization and generalization in pollination systems." *Ecology* in press.

Washitani, I., R. Osawa, H. Namai, and M. Niwa. 1994. "Patterns of female fertility in heterostylous *Primula sieboldii* under severe pollinator limitation." *Journal of Ecology* 82:571–579.

Watanabe, M. E. 1994. "Pollination worries rise as honey bees decline." *Science* 265(16 August):1170.

Weaver, N., and E. Weaver. 1981. "Beekeeping with the stingless bee *Melipona beechei*, by the Yucatecan Maya." *Bee World* 62(1):7–19.

Weiss, S. B., P. Rich, D. D. Murphey, W. H. Calvert, and P. R. Ehrlich. 1991. "Forest canopy structure at overwintering monarch butterfly sites: measurements with hemispherical photography." *Conservation Biology* 5(2): 165–174.

Westrich, P. 1995. "Considering the ecological needs of our native bees: problems of partial habitats." In *The Conservation of Bees*. London: Academic Press.

Whynott, D. 1992. *Following the Bloom: Across America with Migratory Beekeepers*. Boston: Beacon Press.

Williams, C. 1995. "Conserving Europe's bees: why all the buzz?" *Trends in Ecology and Evolution* 10(8):309–310.

Williams, I. H., S. A. Corbett, and J. L. Osborne. 1991. "Beekeeping, wild bees and pollination in the European community." *Bee World* 72:170–180.

Wills, R. T., M. N. Lyons, and D. T. Bell. 1990. "The European honey bee in Western Australian kwongan: foraging preferences and some implications for management." *Proceedings of the Ecological Society of Australia* 16:167–176.

Wilson, E. O. 1992. *The Diversity of Life*. Cambridge: Belknap Press.

Wilson, G. L. 1987. *Buffalo Bird Woman's Garden: Agriculture of the Hidatsa Indians*. St. Paul: Minnesota Historical Society Press.

Wyatt, R. 1983. "Pollination and breeding systems." In L. Real, ed., *Pollination Biology*. New York: Academic Press.

Xerces Society and Smithsonian Institution. 1990. *Butterfly Gardening*. San Francisco: Sierra Club Books.

Yasaka, M., Y. Sunaga, F. Kawasaki, and Y. Konno. 1994. "Effect of forest fragmentation on the fruit set ratio for three perennial herbs." *Japanese Journal of Ecology* 44:1–7.

Yatskievych, G., and R. W. Spellenberg. 1993. "Plant conservation." In Flora of North America Editorial Committee, eds., *Flora of North America: North of Mexico*. Vol. 1. New York: Oxford University Press.

Yurlina, M. E., and S. N. Handel. 1995. "Pollinator activity at an experimental restoration and an adjacent woodland: effect of distance." In *Conserving Europe's Bees*. London: International Bee Research Association.

abiotic pollination. The movement of pollen grains from plant to plant by abiotic vectors. Wind pollination is the most common form of abiotic pollination, although water is used as a vector in a few rare cases by aquatic plants.

abortion. In flowers, an imperfectly developed ovule or fruit that later aborts and falls off the plant.

adaptive radiation. The rapid elaboration of a biological line into various niches as in the case of the Galápagos finches, Hawaiian honeycreepers, or bees of the genus *Perdita*.

Africanized bee (also known as killer bee and bravo bee). A subspecies or race of highly defensive honeybee known as *Apis mellifera scutellata*. These bees, which originated in Africa, were purposefully brought to South America and have now spread into the southwestern United States.

alfalfa leafcutter bee. An introduced Eurasian species, *Megachile rotundata*, belonging to the leafcutter family Megachilidae, that forms the basis for a multimillion-dollar agribusiness for pollinating alfalfa. The bee is easily managed in straws, boards, and styrofoam nesting blocks.

alkali bee. Native North American bee belonging to the genus *Nomia*. These bees often have attractive opalescent bands on the abdomen. One species, *Nomia melanderi,* has been managed as a pollinator of alfalfa in specially constructed alkaline bee beds.

amber. The fossilized resin from various trees, especially the neotropical legume genus *Hymenaea,* but also from certain conifers. Especially abundant in southern Mexico, the Dominican Republic, and the Baltic Sea, amber often contains inclusions including insects and plant debris.

andrenid bee. A member of one of the largest families of bees, the Andrenidae, which are dispersed worldwide but more diverse in north temperate regions, especially on flowers blooming in spring.

**anemophily.** Literally "wind-loving." The pollination of certain flowering plants and gymnosperms by the wind, an abiotic form of pollination.

**angiosperm.** A flowering plant—the dominant plant lifeform today with approximately 246,000 species. Their "double fertilization" and uniquely resistant (to desiccation, fire, and abrasion) seeds may explain their extraordinary evolutionary success since the Cretaceous period.

**anther.** The pollen-containing part of the floral stamens. The anther is usually compartmented into locules—the microsporangium—where the pollen grains are formed.

**anthophorid bee.** The "digger bee" family Anthophoridae, the largest and most diverse family of bees, contains myriad ground-nesting solitary forms as well as parasitic cuckoo bees derived from their ranks. Recently this family has been placed in the honeybee family, Apidae.

**apiculture.** The scientific study of honeybees and their management for increased honey production, beeswax, package bees, queen bees, or commercial rental for pollination services worldwide.

**apid bee.** The common name for any bee in the honeybee family, Apidae, which now includes orchid bees, bumblebees, stingless bees, true honeybees, and the numerous digger bees.

**bee bed.** In the Pacific Northwest of the United States, an innovative approach to bee ranching involving the establishment of artificial bee nesting beds—with upwelling of water and an alkaline crust—for the propagation of alkali bees, *Nomia melanderi,* for alfalfa pollination.

**bee bread.** The vernacular name for pollen combined with nectar or honey and stored in open hexagonal comb cells by honeybees. Since the mixture also contains unique beneficial microbes added by the bees, this can be considered a case of microbial farming.

**bee gardening.** The modern practice of setting out domiciles consisting of drilled boards, hollow straws, and stems to attract native bees to home gardens and public parklands as pollinators of crops and native plants. Sometimes supplemental bees are introduced into these areas.

**beekeeping.** The intentional stewardship of honeybees and stingless bees in hives made of various materials for the harvest of honey, beeswax, and sometimes brood by various cultures. Contrast this husbandry with the earlier honey hunting from which it derived. Beekeeping is at least 5,000 years old.

**biodiversity.** The total spectrum of living variability from gene to species to higher taxonomic groupings and including the ecological interactions, populations, and communities in which they live.

**biogeography.** The study of the geographical distribution of organisms across continents and ocean basins. Biogeography also concerns their habitats along with the biological and historical factors producing the distribution patterns we see today.

**biomechanics.** The mathematical study of the mechanical or physical properties of biological materials—as in the stiffness and resiliency of wood used as cantilever beams or the hydrostatic "water skeletons" of various aquatic animals.

**biophilia.** A term coined by E. O. Wilson to describe humankind's seemingly innate, positive attitudes toward natural diversity and nature itself.

**biota.** The collection of animals, plants, and other organisms occurring together naturally in a certain geographic region.

**bombiculture.** The culture and management of bumblebee colonies (*Bombus* species) to provide pollination services—as in their recent development for the buzz pollination of tomatoes in glasshouses in Europe.

**bulbil.** Plantlet usually sprouting from the flower stalk or base of a mother plant, a form of vegetative reproduction as in century plants in the genus *Agave*.

**bumblebee.** The common name for any bee in the genera *Bombus* or its social parasite *Psithyrus*. These are large, hairy, often black, white and yellow, or reddish truly social bees, with a queen and overlapping generations, having a generally northern distribution.

**bushbaby.** Also known as galagos, this group of seven mammal species is in the family Galagidae from Africa. They are small, agile, arboreal prosimians.

**butterfly gardening.** The intentional planting and arrangement of specialized floral gardens with diverse nectar-producing and larval food plants to attract colorful butterflies. This practice was largely pioneered by Miriam Rothschild in England and is now a popular hobby in Europe and the United States.

**buzz pollination.** A specialized form of pollen harvesting and pollination used by many bees (but not honeybees) to extract pollen from the pored anthers of many flowering plants—for example, blueberries, cranberries, eggplants, tomatoes, and deadly nightshades. About 8 percent of the world's flowering plants exhibit this form of pollination, sometimes called *sonication*.

**calyx.** A Greek word meaning cup. It refers to the usually greenish whorl of sepals that are found at the very base of most flowers. See *sepals*.

**cantharophily.** Literally "beetle-loving." The pollination of certain flowers, usually primitive, by beetles. These flowers often have protein or lipid-rich food bodies.

**Carboniferous.** A geological period that began 365 million years ago and ended 290 million years before the present. A division of the Paleozoic era.

**carpenter bee.** A giant, usually all black, largely tropical bee found in both the New World and Old World. Their strong mandibles are used to excavate galleries in dead wood in which to raise their young. The term mainly refers to the immense genus *Xylocopa,* now placed in the family Apidae.

**cascading extinctions.** The premise that one or several extinctions, especially of key organisms in trophic levels, can lead to a rapid sequence of extinctions of other organisms ecologically linked to the first.

**century plant.** A member of the genus *Agave* with numerous succulent species distributed in the southwestern deserts and mountains of North America. Pollinated by bees in the small spiky forms and by bats, birds, or bees in the candelabra-like forms.

**cheater.** A pollinator that sneaks in the back door—as in the case of nectar-robbing carpenter bees that may slit a tubular corolla to gain access without picking up pollen or effecting pollination.

**chemically induced habitat fragmentation.** The unseen degradation of habitats, especially agricultural or disturbed secondary habitats, by the widespread use of agricultural chemicals and other biocides. To most observers, fragmented habitats may appear intact and even healthy, but are not.

**coevolved.** The idea in evolutionary ecology that certain mutualistic organisms have directed or redirected each other's evolutionary trajectory. Good examples of truly reciprocal coevolution are difficult to find.

**coleoptera.** The largest order of insects, containing almost a million described species of beetles. They are characterized by having the front pair of wings modified into horny sheaths known as the elytra. Many forms, such as scarabs and nitidulids, visit flowers and are effective pollinators.

**colletid bee.** Any of a large family of bees (Colletidae), especially diverse in Australia and other desert habitats, also known as plasterer or membrane bees. They are characterized by having a short, forked proboscis or tongue.

**corbicular.** Refers to the corbiculum ("pollen basket") on the hind legs of female honeybees, bumblebees, stingless bees, and orchid bees. In these social bees the pollen is mixed with nectar and packed into the concavities for transport home.

**Cretaceous.** A geological period lasting 75 million years that began 140 million years ago and ended 65 million years ago with the extinction of the dinosaurs, marine reptiles, and other groups. The earliest flowering plants evolved at the beginning of this period or perhaps somewhat earlier.

**cross-pollination.** The transfer of pollen from the anthers of one plant to a recipient stigma on another plant that may result in fertilization and fruit set. Also known as *outcrossing* or xenogamy.

**Dayang.** In Malaysian folklore, the name for handmaidens of Hitam Manis: the giant honeybee known as *Apis dorsata.*

**DDT.** The abbreviation for the compound known to chemists as dichlorodiphenyltrichloroethan, a highly persistent organochlorine insecticide developed during the 1940s and still widely used outside the United States. Many insects are now resistant to it.

**defaunation.** The process of removing the fauna from a region either intentionally—as in the biogeographic experiments with mangrove islets in Florida—or through the indirect actions of humans.

**diffuse coevolution.** The process by which two or more species with an ecological association evolve more or less together to their mutual benefit.

**dioecious.** Literally "two houses." The separation of male and female flowers on different plants as in jojoba *(Simmondsia chinensis).*

**dipterocarp.** The common name for a tall canopy emergent tree in the Old World rainforest of Malaysia and other parts of Asia. They belong to the botanical family Dipterocarpaceae, as do most of the tree species in these Asian rainforests.

**directional selection.** The nonrandom differential reproduction or spread of various genotypes in a population. Here selection is "directed" by various forces toward one extreme.

**domicile.** A material modified, usually by drilling holes, to serve as a nest site for leafcutter and other bees under a pollination management scheme. The domiciles may be paper straws, drilled wooden boards, cardboard, styrofoam, or pithy stems.

**endangered phenomenon.** A phenomenon that is threatened with extinction, usually by human activity.

**endangered species.** A species that is in danger of (global) extinction.

**endemism.** The process whereby a species is restricted to a small geographic area. "Endemic" species, therefore originated or evolved in that area.

**epiphyte.** A plant that grows upon another plant, especially in tropical rainforests. Epiphytes are not parasitic. Examples include many orchids, bromeliads, and ferns.

**euglossines.** A tribe of brilliantly colored, shiny, metallic bees found only in the New World tropics. The males collect volatiles from orchids and other botanical sources and form leks. The females often visit flowers with pored anthers and sonicate them.

**extirpation.** The elimination of a population from a locality either by human activity or natural causes.

**facultative mutualism.** A mutualistic (beneficial to both partners) relationship that is not obligate. That is, the partners need not enter into an ecological "pact" in order for them to prosper and survive to leave more offspring.

**fauna.** The animal assemblage found living in a given region. See *flora* for plants.

**fertilization.** The penetration of the egg cell membrane by an individual sperm cell at the time of conception. Once fertilized, the combination of egg plus sperm is known as a zygote, which undergoes further embryogenesis and cell divisions in the next few hours and days.

**filament.** The usually thin supportive stalk that bears the pollen-containing anthers in a flower. Together the filament and anther make up the stamen, the male reproductive organ of a flower.

**flabellum.** On an insect's pair of antennae ("feelers") this is the long wandlike group of segments that bear the sensory cells. They allow the bee, for example, to detect the presence of water molecules and odors in the air.

**flora.** The plants found together in a geographic region. See *fauna* for animals.

**floral reward.** A diverse array of attractants and often nutritious food present in flowers to invite and lengthen visits by floral visitors. The major substances used are pollen, nectar, floral oils, and food bodies.

**floral visitor.** Any animal that visits a flower to find food, shelter, a mate, or simply to rest. Such visitors should not be confused with legitimate pollinators that effect pollination and subsequent fertilization of ovules.

**flying fox.** Large—up to 4-foot wingspan—member of the bat family Pteropodidae from the Old World tropics. Flying foxes are important pollinators and seed dispersers of rainforest plants.

**food body.** Specialized protein and lipid-rich plant tissue often found in primitive flowers such as western spicebush. Food bodies serve as rewards and prolong the stay of beetle pollinators within such flowers.

**fruit set.** Fruit is set by blossoms when the ovules are fertilized and the plant has enough energy and water reserves to develop and ripen the fruit. Fruit set is enhanced by pollination.

**gecko.** Any member of the lizard family Geckonidae. One insular species from offshore islands near New Zealand has modified throat scales and is a pollinator—perhaps the only true example of pollination by reptiles.

**gene pool.** The alleles for all genes within a freely interbreeding population of organisms drawn upon by plant breeders.

**generalist.** A pollinator that visits a wide variety of flowers for nectar and pollen during its lifetime. A generalist blossom is one that is quite open and can be visited and pollinated by unrelated groups of animals.

**genetic recombination.** In sexually reproducing organisms, this is the shuffling of genes resulting in new combinations during the "sexual lottery" in which certain genes are carried by the lucky sperm to an egg. During the elaborate process of meiosis, one set of chromosomes (with its set of genes) is donated by the father and one set by the mother. This can result in novel genotypes in the offspring that can often allow them to survive and reproduce in diferent or otherwise harsh environments. Since no two individuals are genetically identical, ever-new gene combinations are created by sexuality.

**genetic variation.** The various genes and their relative frequencies within a population of organisms.

**gymnosperm.** An ancient division of seed-bearing vascular plants from the Devonian period to recent times. They reproduce by clusters of cones (strobili) and are largely pollinated by the wind.

**habitat corridor.** A linear band or series of "stepping stones" of undisturbed or secondary habitat connecting two protected regions—as in proposed park and wildlife preserve designs allowing animals to cross from one habitat to the other to find food and shelter.

**habitat fragmentation.** The division of natural ecosystems into patchwork habitats due to land conversion for agriculture, forestry, and urbanization. Such jigsaw pieces of land are eventually too small and too widely scattered to support the full complement of former species, many of which may then become extirpated.

**hand pollination.** The human pollination of flowers, usually crop plants or those growing in research plots, with fresh or stored pollen. Some endangered native plants now have to be pollinated by hand to set any fruits and seeds.

**hawkmoth.** Large group of moths in the family Sphingidae with tongues (proboscides) often longer than their bodies. Often known as sphinx or sphingid moths, these energetic insects specialize on the rich nectar rewards within fragrant, usually white, tubular blossoms opening at night. The familiar tobacco/tomato hornworm is a member of this group.

**heterozygosity.** A measure of the genetic diversity (at the gene level) in a natural population. It takes into account the number of different gene loci (sites) on the chromosomes for different individuals.

**honey hunting.** The ancient practice of raiding the colonies of social honeybees, or stingless bees, for honey, beeswax, and larvae for human food and other purposes. Depicted in cave paintings and petroglyphs in many parts of the world, honey hunting predates beekeeping.

**honey possum.** A family of small mammals, the Tarsipedidae, found in southwestern Australia. They have a long protrusible tongue for feeding on nectar and pollen and are excellent pollinators.

**honeycreeper.** A family of 23 species, the Drepanididae, of small tree-living perching birds found only in the forests of the Hawaiian Islands.

**hummingbird.** All the members of the New World bird family known as the Trochilidae. Extreme specialization for feeding on nectar from tubular, often reddish blossoms is found in the New World family Trochilidae, with about 340 colorful species, many endangered. Many are migratory along nectar corridors, some may use a traplining feeding strategy, and some are important pollinators. Others are robbers or cheaters.

**hybridization.** Reproduction between two related species that results in the formation of viable progeny usually intermediate in character between the parental types.

**hydrophily.** Pollination by fresh or salt water, as in the case of eelgrass. Since pollen is usually damaged by immersion in water, this form of pollination is rare.

**hymenoptera.** The second largest order of insects. Includes such groups as sawflies, bees, ants, and solitary and social wasps. The name ("membrane wing") refers to their two pairs of diaphanous wings. They have a complete metamorphosis extending from egg to larva through pupa and the resulting adult, or imago.

**inbreeding.** Sexual reproduction that involves the interbreeding of closely related individual plants through self-pollination and often backcrosses.

**insecticide.** Any of a number of commonly available, usually organically synthesized, agrichemicals designed to kill insects that are economic pests. Most of them are non-target-specific, however, and kill or debilitate countless beneficial animals including pollinating bees and other insects.

**jimsonweed.** The common name for the genus *Datura,* with several species found in the American southwestern deserts. Jimsonweed produces large, sweetly scented white flowers that bloom at night and are effectively pollinated by hawkmoths.

**keystone mutualist.** A plant or animal whose importance in a community is inordinately tied to other plants and animals. When removed from a system, keystone mutualists can cause a cascade of linked extinctions. Fig trees in the neotropics are an example.

**leafcutter bee.** Any bee in the large and diverse family Megachilidae. These bees nest in abandoned insect burrows in logs or other holes, sometimes in the ground. They carry their pollen dry in a brush of hairs beneath the abdomen. The best known of this group, the alfalfa leafcutter bee *(Megachile rotundata),* is a managed pollinator of alfalfa and other crops.

**lek.** A form of display used by male animals—as in the case of certain birds like the sage grouse—in which males with exaggerated plumage strut and vocalize in the presence of females who choose a mate based on his performance. Although leks are rare among insects, some carpenter bee males display in nonflowering trees and shrubs that have no food to offer visiting females as prospective mates.

**lepidoptera.** The insect order that contains the butterflies and moths: insects with two pairs of wings covered in scales. Although butterflies are more noticed, moths vastly outnumber them and represent more important pollinators.

**linked extinctions.** Extinctions of plants or animals may be "linked" in the sense that when one organism goes extinct, it may be living in a mutualistic relationship with others in a food web and the original event may cause other lifeforms to go extinct also.

**mason bee.** The common name given to any bee in the genus *Osmia* and belonging to the leafcutter bee family. They construct nests with leaf pieces and mud. One species, *O. lignaria,* the blue orchard bee, is easily managed as an important pollinator of orchard crops.

**meliponiculture.** The ritualized keeping of stingless bees (meliponines), especially in a traditional context by Mayan Indians in parts of Mexico and Central America. These bees are usually kept in hollow logs and periodically raided to obtain honey and wax.

**meliponines.** The stingless bees belonging to the genera *Melipona* and *Trigona,* which occur in both the New World and Old World tropics. They are highly social and live in populous perennial colonies.

**melittophily.** The scientific study of all bees—as opposed to apiculture, the management of honeybees in the genus *Apis.*

**membrane bee.** Any bee belonging to the family Colletidae. Also known as plasterer bees, membrane bees secrete a polyester membrane to line the cells of their underground nests. They are most abundant in Australia.

**mescal.** A distilled bootleg liquor made in Mexico from cooked and fermented century plants in the genus *Agave.* In Sonora, Mexico, mescal *"bacanora"* is to tequila what moonshine is to whiskey.

**mess-and-soil pollination.** A type of pollination usually occurring in primitive flowering plants and effected by beetles—so named because the beetles blunder around the flowers chewing on floral parts, eating pollen, and defecating.

**migratory beekeeping.** A form of beekeeping practiced in the United States and other technologically advanced countries with large beekeeping industries. Basically, beekeepers use trucks to transport their colonies to "follow the bloom" and custom-pollinate agricultural crops such as almonds in California.

**monoecious.** Literally "one house." A reproductive system in which both male and female blossoms are borne on the same plant—as in the squashes, gourds, and pumpkins, for example.

**monolecty.** The pollen-collecting behavior of certain solitary bee species that, throughout their range and year to year, specialize on the pollen of just one species or related congeners. See *oligolecty* and *polylecty*.

**muscid.** Any member fo the large fly family Muscidae in the order Diptera.

**mutualism.** A type of symbiosis in which all partners derive benefits from the association.

**myophily.** Literally "fly-loving plants." The pollination of flowers by flies.

**nectar.** A watery floral secretion containing sugars, amino acids, lipids, and often other compounds such as antioxidants. Nectar varies greatly in its sugar concentration: bees prefer nectar in the 30 to 50 percent range.

**nectar corridor.** A series of different plants offering abundant nectar seasonally along an annual migration pathway—as in the case of century plant and cactus offerings to bats in the southwestern United States and Mexico.

**nectar robber.** A floral visitor that "burglarizes" blossoms by taking nectar by forceful entry without effecting pollination.

**nectary.** A specialized region of floral tissue, usually at the base of the innermost floral tube, where nectar, a mixture of sugars and amino acids in water, is secreted. The nectar solution forms pools in this region and pollinators drink from it.

**non-Apis bee.** Casual term employed by melittologists for any bee not belonging to the honeybee genus.

**obligately outcrossing.** A flowering plant species that, for reasons of physiological incompatibility, must receive and donate pollen in the form of cross-pollination. It cannot pollinate itself between flowers on the same plant. Greater levels of seed set are achieved with outcrossing, as well as increased genetic recombination (mixing of genes).

**obligate mutualism.** A mutalistic relationship between partners who cannot survive outside the relationship. Contrast with *facultative mutualism*.

**oligolecty.** The pollen-collecting behavior of certain solitary bees that specialize on a few related flowers for their pollen needs. Their floral constancy extends throughout the distributional range of their host plants, the same species of which are used year after year. See *monolecty* and *polylecty*.

**orchid bee.** Any of a group of extremely specialized apid bees that pollinate certain tribes and genera in the immense orchid family. Known also as euglossines, orchid bees are shiny metallic, usually green or blue in color, and the males are highly specialized to visit orchids and harvest their floral scents. Species of some orchid bee males are thought to display to females in mating leks within clearings in tropical forests. Orchid bees occur only in the New World.

**organochlorine.** A large family of insecticidal chemicals containing chlorine atoms. DDT, chlordane, and heptachlor are common organochlorines.

**ornithophily.** Pollination of bird-adapted flowers by perching and hovering birds. Usually these flowers are not only large and colorful but also sturdily built so the birds' probing bills and feet do not damage the ovules within the ovaries.

**outcrossing.** The chief means of reproduction by flowering plants. Such plants can only be fertilized by pollen from other plants.

**ovary.** Lower portion of the floral pistil containing the ovules that when fertilized will become the seeds in the fruits of flowering plants.

**ovule.** Structure in the female portion of the flower that becomes the seed.

**paleobotany.** The study of long-dead fossilized plants. Paleobotany usually includes their formal naming and classification along with research on their phylogeny, paleoecology, or even chemical residues.

**Paleocene.** Epoch in the Cenozoic era that began 65 million years ago at the end of the Cretaceous period and ended 11 million years later.

**paleoecology.** The study of the fossil environments: deciphering what conditions were like when those plants and animals were alive.

**pedicel.** The very base of a flower: the slender stalk that supports the flower and attachs it to the parent plant.

**perianth.** A botanical term that collectively refers to the whorls of petals in a typical flower. This is the floral tissue that is brightly colored and produces the floral scents.

**petals.** The showy, often colorful array of plant organs within flowers that have evolved as "billboards" and perfume dispensers to advertise their presence to pollinators and guide them to hidden nectar deposits.

**pin flower.** Specialized form of different flowers on the same plant. Some have long pistils (these are the "pin" flowers) while others ("thrum" flowers) have very short pistils bearing the receptive stigmas. Compare with *thrum flower.*

**pistil.** The female reproductive organ of a flower that is further subdivided into ovary, style, and stigma.

**plant/pollinator landscape.** A term introduced by Judith Bronstein to describe a situation in which pollinators and their floral plant hosts have evolved in a mutualistic way over long periods of time.

**pollen.** Microscopic particles in anther locules that contain the male sperm nuclei. The pollen grain, or microspore, consists of a highly resistant, usually sculpted, shell and an interior cytoplasm.

**pollination.** The process of moving pollen from the anthers of one flower to the stigma of another. Equally vital processes of fertilization and seed set follow from pollination. Pollination can be effected either by abiotic means such as wind and water or by animals such as bats and bees.

**pollination ecology.** The study of the ecological and evolutionary relationships involved in the pollination process.

**pollination services.** Pollination acts performed by all the various animals that dependably visit certain species of flowering plants.

**pollination syndrome.** An old concept, dating back to the early works of Europeans including Paul Knuth and Hermann Mueller, suggesting that the floral visitors a flower receives (hummingbird, bat, fly, butterfly, moth, bee, beetle) can be predicted from the flower's suite of morphological characteristics and rewards. Although subject to misapplication, the concept is still of great value in teaching about pollination.

**pollinator.** Any animal that not only visits a flowering plant but effects pollination, leading to fertilization and later fruit and seed set. Not all floral visitors are legitimate pollinators, however, so that long lists of floral visitors are of dubious use in determining who actually performs pollination.

**pollinator garden.** A garden specifically designed to attract butterflies, hummingbirds, or bees. The modern pollinator garden dates back to the *Schulgartens* (school garden) of the seventeenth century.

**pollinator-limited.** A system in which pollinators—and therefore pollen grains—are in short supply and thus limit the amount of fruits and seeds that can be produced by a local plant population.

**pollinator pool.** All of the pollinators that visit a certain plant in a given area.

**pollinator syndrome.** A system used by many floral biologists during the middle of the twentieth century to describe certain groups of flowers classified according to their physical attributes (shape, size, open or tubular, and so on) as well as their color, odors, and types and amounts of floral rewards offered. Thus deep-throated red flowers with no scent and abundant nectar are most often pollinated by hummingbirds in the New World tropics. Therefore, such blossoms are said to be hummingbird-pollinated and belong to this pollinator syndrome. Similarly, one can describe floral "guilds" that are adapted by pollination by flies, beetles, birds, butterflies, moths, and bats.

**pollinium (pl. pollinia).** A mass of pollen grains united into a single dispersal unit for more efficient transport by a specialized pollinator. Pollinia are routinely found in the milkweed and orchid families. This pollen is usually quite safe from ingestion by pollinators.

**polylecty.** The pollen-collecting and feeding habits of social bees and certain solitary bees that routinely gather pollen from unrelated plant families and genera in one area and over a long period of time. One of the most polylectic species is the introduced honeybee, *Apis mellifera,* fot it the broadest pollen diet. See *monolecty* and *oligolecty.*

**proboscis, (pl. proboscides).** The insect's tongue apparatus. It can be long and coiled, as in butterflies, or quite short and spongelike as in flies.

**provision mass.** The amassed mixture of pollen, nectar, and often bodily secretions from female bees used as the complete diet for solitary bees. Such bees are said to be mass-provisioned because their larvae receive all the food they require, from egg to adult, at one time from their mother.

**reciprocal coevolution.** The most extreme type of mutualism: each of the partners evolves with its evolutionary "dance partner"—for example, the long-legged bee genus *Rediviva* and its twin-spur floral host genus *Diascia* from southern Africa. Neither can live without the other, and apparently both have directed each other's evolutionary trajectory.

**restoration ecology.** The science and process of using plants and animals to restore a human-degraded site to its former condition. Examples include reforestation of timberland or mine tailings.

**rodenticide.** A biocide designed for the killing of rodents.

**safe sites.** In seed ecology, the final resting spot for a seed where it can finally germinate and grow into a seedling or mature plant. It is safe from seed predators and doesn't succumb to harsh enviromental conditions. Similarly, we can think of "safe sites" on bees where they cannot groom off the pollen grains deposited by certain flowers.

**scramble competition.** A free-for-all competition in which competitors scramble for limited available food or other resources.

**seed bank.** The total buried reserve of all the seeds, new and old, from all seed plants growing in one area—a safety mechanism to ensure that new recruits of all the species can germinate in future years.

**seed shadow.** The dispersal pattern of seeds disseminated by a plant and scattered in various directions and distances by different seed-dispersal agents.

**self-fertile.** Flowering plants that can set fully viable seeds when pollinated with their own pollen. The same as self-pollinated.

**self-incompatible.** Flowering plants that are incapable of pollinating themselves and rely on genetically distinct pollen from distant plants or nonrelatives to form fruits and seeds.

**sepal.** A floral part that is usually divided into a whorl of petal-like green appendages at the base of the flower. Collectively they form the calyx, which serves a protective function. See *calyx.*

**sequential mutualism.** When plants growing in the same area overlap in blooming period and are pollinated by a group of coadapted mutualists—as in the case of red tubular or blue alpine meadow flowers that are pollinated by different sets of hummingbird species as the season progresses.

**Silurian.** The portion of the Paleozoic era that began 441 million years ago and lasted for 28 million years.

**social bee.** Social bees are those that live together in a communal nest and often share foraging or nest duties. The highest form of sociality involves an overlap of generations, such that a mother bee will share a nest with her offspring. Also known as eusocial, these bees—including the familiar honeybees, bumblebees, and certain sweat bees—have overlapping generations and distinct worker, male, and queen castes, sometimes along with a division of labor. See *solitary bee.*

**solitary bee.** Most of the world's 40,000 or so species of bees are solitary. That is, each female mates and then begins construction of underground (or within woody stems) nests that branch and end in smooth-walled cells. These cells are filled with a mixture of nectar and pollen that provide all the food required by the young larva to complete its development into an adult bee. See *social bee.*

**sonication.** The act of releasing pollen from a flower with pored anthers by a female bee using strong vibrations produced by shivering the thoracic flight muscles. Also known as buzz pollination, sonication occurs in blueberries, cranberries, eggplants, kiwi fruits, and tomatoes.

**spore.** A reproductive cell capable of growing into an adult plant without fusing with another sex cell.

**squash bee.** A group of anthophorid ground-nesting bees in the genera *Peponapis* and *Xenoglossa* that depend on plants in the genus *Curcurbita* for all their pollen and nectar food requirements.

**stamen.** The male structure bearing the pollen grains in a flower. It is made up of two parts: a long slender stalk, or filament, and the chambered pollen-producing anther that sits atop it.

**stigma (pl. stigmas).** The glandular female receptive portion at the end of the carpel/pistil where pollen grains land and germinate, sending down pollen tubes and sex cells.

**Stil.** A German term meaning style; introduced by Stefan Vogel, an eminent pollination biologist.

**stingless bee.** Group of social bees from the New World and Old World tropics that form large perennial nests and have well-established castes. Stingless bees have only a vestigial sting apparatus but often defend their colonies with aggressive attacks including biting the intruding animal (usually a large mammal).

**stochastic.** Random; unpredictable.

**style.** The middle "connective" portion of a floral pistil uniting the stigma above and ovary below.

**subspecies.** A taxonomic subdivision of a species that is considered a distinct geographical race.

**subtropical thornscrub.** A vegetational association just outside the true tropics that is drought-deciduous and characterized by having large numbers of thorny tree and shrub species.

**sugar glider.** Any of a group of arboreal native Australian marsupials adapted for living off the sweet exudates of trees as well as the nectar and pollen of flowering plants. The mammals are pollinators of some of these plants.

**sweat bee.** Any member of the bee family Halictidae. These bees can be highly social with well-developed castes. Often they are attracted to sweating humans where they imbibe the perspiration for moisture.

**symbiosis.** Two or more dissimilar organisms living together. The association may benefit only one or perhaps all of the so-called symbionts.

**syrphid.** Any member of the large fly family Syrphidae. These brightly colored yellow and black flies mimic bees and wasps in their darting flight and eagerly visit blossoms to lap up nectar and eat pollen. Many of them are important pollinators.

**tamarin.** Along with the marmoset, this mammal forms the family Callitrichidae with 14 species of small insectivorous primates. They occur in the neotropical forested regions.

**taxonomy.** The scientific classification and naming of organisms.

**tepal.** A whorl in the perianth of a flower that is not differentiated into distinct petals or sepals—often seen in primitive angiosperms.

**threatened species.** As used in U.S. federal law, a plant or animal species that is likely to become endangered in the near future.

**thrip.** Very small flying insect with fringed wings and rasping mouthparts. Usually detrimental to flowers by eating pollen, thrips are important pollinators of many tree species in certain forests of Southeast Asia.

**thrum flower.** Some flowers were noted by Darwin and other early workers to belong to two types although they were produced on the same plant—a method to promote outcrossing and enhance pollen donation between distant plants. In one of the most common types, there are "pin and thrum" blossoms. The thrum flowers have short stigmas often hidden deep within the flower. Pollen can only be successfully donated (and germinate) on a stigma when it goes from pin to thrum or vice versa. See *pin flower.*

**traplining.** A feeding strategy in which certain birds and insects follow a "trapline" of blooming plants in a set order on a daily basis. These animals are familiar with the distribution and flowering status of plants on their trapline.

**trap nest.** Block of drilled wood or bundle of straws or hollow stems attached to tree trunks or buildings to provide bee nesting sites near crop or garden plants for pollination.

**tripped flower.** In certain legumes, such as alfalfa, floral parts are held shut until visited by a bee that forces the blossom open. This process often results in a sudden upswing of the anthers, dusting the underside of the bee. By observing the ratio of tripped to untripped blossoms in a field, farmers can tell if their crop is being adequately pollinated.

**vector.** Any biological or abiotic (e.g., wind or water) agent which carries something around in a directed fashion. In our case, pollinators "vector" or mine pollen grains around the environment and from flower to flower.

**vegetative reproduction.** An alternative form of plant reproduction in which a piece of the parental plant or an offset (see *bulbil*) grows into a genetically identical clone yet distinctly separate plant.

**vestigial organ.** A plant or animal organ that appears to have had an important function at one time but is no longer or rarely needed.

**vulture bee.** The informal name given to a small group of stingless bees in the genus *Trigona* by David Roubik. So far as we know, they are the only example of carrion-feeding bees.

**white-eye.** Flower-visiting bird in the family Zosteropidae with about 80 species in forested regions from Africa to New Zealand. White-eyes have a grooved and frilled tongue and feed on insects, nectar, and fruit.

**wind-pollinated.** Plants that depend on the wind to carry their pollen grains from plant to plant. Pine trees and jojoba are examples.

**yellow rain.** The unique mass defecation flight of the giant Asian honeybee, *Apis dorsata*.

**zygote.** The result of the fusion of a sperm sex cell with an unfertilized egg. Following fertilization, this zygote begins to undergo cellular division and grows into a recognizable embryo in a matter of hours or days.

A Call For a National
Policy on Pollination

As scientists and educators, food producers and consumers, we are concerned that a basic fact of life—our dependence on the link between plants and their pollinators—is poorly understood by policymakers, the public, land managers, and the food industry. We have joined together to urge further appreciation and appropriate action regarding the following principles:

1. *The Future of Our Farms Depends on Pollination.* The pollination of plants, which is often mediated by animals, is necessary for seed set, fruit yields, and reproduction of most food crops. Animal pollinators also serve a diversity of dominant trees and herbs in wildlands. Nevertheless, pollination remains one of the weakest links in our understanding of how ecosystems function and how crop yields can be assured. Pollination ecology should be taught in every agricultural program and land grant college in the United States.

2. *We Need to Appreciate the Benefits that a Diversity of Pollinators Provides.* Pollination services are provided by a diversity of animals in addition to the domestic honeybee. And yet the "free" pollination services provided to food and forest crops seldom enter into government statistics on the value of protecting wild species or the "costs" of maintaining agricultural yields. A complete inventory of pollinators of crops and keystone plant species in wildlands should be compiled by the National Biological Service and USDA-ARS and their counterparts in other countries.

3. *Honeybees Are in Decline.* Since 1990, U.S. beekeepers have lost one-fifth of all their domestic managed honeybee colonies due to the arrival of Africanized honeybees and abandonment of hives, the spread of parasitic mites and diseases, and the withdrawal of honey subsidies that formerly helped support the industry. Wild pollinators must now take up some of this slack in pollinating crops. Collectively, wild and domestic bees provide pollination services 40 to 50 times more valuable than the market price of all honey produced in the United States. The USDA, Mexico's SARH, and Agriculture Canada should invest more in their non-*Apis* bee programs and provide support in order to stabilize the honeybee industry and encourage diversification in pollination industries.

4. *All Pollinators Require Protection from Toxins and Land Degradation.* Maintenance of wild pollinators for crops requires habitat set-asides or greenbelts near agricultural fields. Since wild pollinators are often more vulnerable to pesticides and herbicides than are domestic honeybees, the use of toxic chemicals must be carefully controlled near their nesting and foraging sites. Those who apply pesticides should be better trained in monitoring the health of pollinator populations. When given a choice, they should use insecticides that are known to be less toxic for bees.

5. *Habitat Fragmentation Is A Major Threat to Pollinators.* Although its effects on native pollinators are not completely understood, habitat fragmentation may be reducing their populations due to loss of nesting habitats, indiscriminate use of rangeland and agricultural chemicals, and elimination of host plants and nectar sources. As habitat patches ("islands") become smaller, they may become insufficient to support pollinators through the mix of plants they require. Populations of pollinators should be closely monitored and habitat fragmentation trends mapped in order to determine specific causes of pollinator decline and convince land-use planners of the need to zone greenbelts and corridors to ensure pollinator survival.

6. *Fewer Pollinators Ultimately Means Fewer Plants.* Where habitat fragmentation and pesticides have reduced the populations of pollinators, plants will eventually suffer low reproductive success. The pollination ecology of many plants has barely been studied, even though it may be critical to keeping some of these plants from extinction. We must earmark support for studying these interactions and for including pollinator nesting and foraging areas in critical habitat designations.

7. *Endangered Species Protection Need Not Be Incompatible with Food Protection.* Pesticides and herbicides must not be sprayed in the immediate vicinity of endangered plants, rare pollinators, and their habitats. Nevertheless, the "spraying setback distances" being set by the EPA and USFWS have been arbitrarily determined without detailed knowledge of specific

plant/pollinator relationships. Such arbitrary determinations not only frustrate ranchers and farmers, but may fail to protect the species they are meant to conserve. Before implementing such setbacks, on-site determinations need to be made by pollination ecologists familiar with the species involved.

8. *Plants and Pollinators Both Need Protected Habitats.* In a few cases, pollinators have declined to the extent that the economically important plants they formerly served are suffering reduced seed set and fruit yields. In other cases, the decline of certain plants has triggered the decline of pollinators that specialize on them. These reciprocities deserve special attention and support to reverse these trends and restore the relationships. When necessary to keep them from extinction, critical habitat should be set aside for both plants and their pollinators.

9. *Migratory Pollinators May Require International Protection.* Although the reproduction of plants is occasionally reduced by lack of pollinators, this condition has been aggravated by human activity in the landscapes they inhabit and the severing of migratory corridors. Scientists must carefully monitor plant/pollinator changes and assess global trends.

10. *Pollination Is a Threatened Ecological Service.* The loss of biological diversity means more than simply counting the declining number of species. It also implies the extinction of relationships or disruption of ecological processes, such as pollination, upon which we all depend. One in every three mouthfuls of food we eat depends on pollination by bees and other animals to reach our kitchen tables. And yet, thousands of pollinating animal species are globally endangered. Additional habitat conservation, monitoring, research, and ecological restoration will be required to reverse these far-reaching global trends.

The above "Ten Point Plan," a call to establish a national pollination and pollinator protection policy, was discussed and adopted by the Forgotten Pollinators advisory board during the first workshop conducted at the Arizona-Sonora Desert Museum, Tucson, Arizona during the spring of 1995. The following advisory board members adopted the plan: Melody Allen, Dr. Peter Bernhardt, Dr. Ron Bitner, Dr. Judith Bronstein, Dr. Stephen Buchmann, Dr. James Cane, Elizabeth Donnelly, Dr. Peter Feinsinger, Charlotte Fox, Mrill Ingram, Dr. Peter Kevan, Dr. Andrew Matheson, Ing. Sergio Medellin-Morales, Dr. Gary Paul Nabhan, Dr. Robert Michael Pyle, Steve Prchal, Dr. Beverly Rathke, Dr. Vince Tepedino, Dr. Philip Torchio, and Steve Walker.

APPENDIX 2

# Pollinators of the Major Crop Plants

| CROP SPECIES[a] | KNOWN POLLINATORS[b] |
| --- | --- |
| Wheats (2) (*Triticum* spp.) | wind |
| Rices (2) (*Oryza* spp.) | wind |
| Maize *(Zea mays)* | wind |
| Sorghum *(Sorghum bicolor)* | wind |
| Millets (5+) *(Echinochloa, Eleusine,* *Panicum, Pennisetum, Setaria)* | wind, bee, insect |
| Rye *(Secale cereale)* | wind |
| Barley *(Hordenum vulgare)* | wind |
| Oats *(Avena sativa)* | wind |
| Fonio *(Digitaria exilis)* | bee |
| Quinoa *(Chenopodium quinoa)* | wind |
| Potato *(Solanum tuberosum)* | bee *(Bombus)* |
| Cassava *(Manihot esculenta)* | bee |
| Yams (5+) *(Dioscorea* spp.) | bee, beetle, fly |
| Sweet potato *(Ipomaea batatas)* | bee, *(Melitoma, Ancylosceius)* |
| Taro *(Colocasia esculenta)* | fly |
| Yautia *(Xanthosoma sagittifolium)* | bee |
| Sugarcane *(Saccharum offinarum)* | bee, thrips |
| Sugar beet *(Beta vulgaris)* | wind, small insects |
| Soybean *(Glycine max)* | bee |
| Groundnut *(Arachis hypogaea)* | bee *(Megachile, Liptriches)* |

| CROP SPECIES[a] | KNOWN POLLINATORS[b] |
|---|---|
| Beans (5+) *(Phaseolus, Lablab, Vigna)* | bee, thrip; bee, thrip; bee; bee |
| Cowpea *(Vigna unguiculata)* | bee |
| Pea *(Pisum sativum)* | bee, thrips |
| Pigeonpea *(Cajanus cajan)* | bee *(Chalicodium, Xylo)* |
| Chickpea *(Cicer arietinum)* | bee |
| Broadbean *(Vicia faba)* | bee |
| Lentil *(Lens culinaris)* | insect, bee |
| Coconut *(Cocos nucifera)* | bee, wind, fly, bat *(Eonycteris, Macroglossus, Pteropus)* |
| Sunflower *(Helianthus annuus)* | bee *(Apis, Melitoma,)* fly |
| Oil palm *(Elaeis guineensis)* | beetle *(Elaeidobius,)* insect, bee |
| Cottonseed *(Gossypium barbadense)* | bee |
| Olive *(Olea europaea)* | wind |
| Rapeseed *(Brassica napus)* | bee *(Apis)* |
| Sesame *(Sesamum orientale)* | bee, fly, wasp |
| Melonseed *(Cucumis* spp.*)* | bee |
| Karite nut *(Vitellari paradoxa)* | bee |
| Almond *(Prunus dulcis)* | bee |
| Filbert *(Corylus avellana)* | wind |
| Mustard *(Brassica juncea)* | bee *(Apis)* |
| Safflower *(Carthamus tinctoria)* | insect, bee *(Apis)* |
| Walnut *(Juglans major)* | wind |
| Brazilnut *(Bertholletia excelsa)* | bee *(Euglossa, Epicharis)* |
| Pistachio *(Pistachia vera)* | wind |
| Tomato *(Lycopersicon esculentum)* | bee *(Bombus, Exomalopsis)* |
| Cabbage *(Brassica oleracea)* | bee *(Apis)* |
| Onions (2) *(Allium* spp.*)* | bee *(Apis),* fly |
| Cabbage *(Brassica oleracea)* | bee *(Apis)* |
| Carrot *(Daucus carota)* | fly, bee *(Apis)* |
| Cucumber (2) *(Cucumis* spp.*)* | bee *(Apis)* |
| Pumpkins (4) *(Cucurbita* spp.*)* | bee *(Apis, Xenoglossa, Peponapis)* |
| Lettuce *(Lactuca sativa)* | bee, small insects |
| Chili pepper *(Capsicum annuum)* | bee |
| Eggplant *(Solanum melanogena)* | bee |
| Garlic (2+) *(Allium* spp.*)* | bee *(Apis),* fly |
| Spinach *(Spinacia oleracea)* | wind, insect |
| Artichoke *(Cynaria scolymus)* | bee *( Bombus)* |
| Bananas (2+) *(Musa* spp.*)* | bird, bat *(Cynopterus, Macroglossus)* |

| CROP SPECIES[a] | KNOWN POLLINATORS[b] |
|---|---|
| Orange *(Citrus sinensis)* | bee *(Apis)* |
| Watermelon *(Citrullus lanatus)* | bee *(Apis)* |
| Date *(Phoenix dactylifera)* | wind, bee? |
| Avocado *(Persea americana)* | bee, fly, bat *(Pteropus)* |
| Mango *(Mangifera indica)* | bee, fly, bat *(Pteropus)* |
| Pineapple *(Ananas comosus)* | bird |
| Tangerine *(Citrus reticulata)* | bee *(Apis)* |
| Lemon/Lime (2) *(Citrus* spp.) | bee *(Apis)* |
| Grapefruit (2) *(Citrus* spp.) | bee *(Apis)* |
| Melon *(Cucumis melo)* | bee *(Apis)* |
| Papaya *(Carica papaya)* | moth, bird, bee |
| Pear *(Pyrus communis)* | bee, fly |
| Peach *(Prunus persica)* | bee |
| Plum *(Prunus domestica)* | bee |
| Fig *(Ficus carica)* | wasp *(Blastocephala, Ceratopogonidae)* |
| Strawberry *(Fragaria* × *ananassa)* | bee |
| Apricot *(Prunus armenica)* | bee |
| Cherry *(Prunus avium)* | bee |
| Currant (2+) *(Ribes* spp.) | bee, wasp, fly |
| Allspice *(Pimienta dioica)* | insect, bee |
| Star anise *(Illicium verum)* | bee, wasp, fly |
| Cardamon *(Elettaria cardamomum)* | bee *(Apis, Amegilla)* |
| Pepper *(Piper nigrum)* | bee, fly |
| Cocoa *(Theobroma cacao)* | fly *(Forcipomyia, Lasiohela)* |
| Coffee (2) *(Coffea* spp.) | fly, bee *(Apis, Melitoma)* |
| Mate *(Ilex paraguariensis)* | bee |
| Tea *(Camellia sinensis)* | insect, fly, bee |

[a] This inventory of the crop commodities most important in providing calories, proteins, and fats to human diets is modified from R. Prescott-Allen and C. Prescott-Allen 1990. "How many plants feed the world?" *Conservation Biology* 4(4):365–374.

[b] Pollinators are from Roubik (1995), Free (1993), and personal observations and USDA Advisory Committees.

# Conservation and Research Organizations

Perhaps the best single reference on conservation-related activities is the *Conservation Directory*, updated annually and published by the National Wildlife Federation. This comprehensive directory contains thousands of listings on local, national, and international conservation organizations, conservation publications, and the names of leaders in many conservation areas. Other exhaustive references of interest are *The New Complete Guide to Environmental Careers* (Island Press, 1993) and *Environmental Profiles: A Global Guide to Projects and People* (Garland Publishing, 1993). The following list provides addresses and a brief description of activities for some major environmental organizations and resources.

American Association of Zoological Parks and Aquariums
Oglesbay Park
Wheeling, WV 26003
   *This national association is dedicated to promoting the responsible care and exhibition of animals by zoos. It is a strong advocacy group for the preservation of captive wildlife species.*

Arizona-Sonora Desert Museum
2021 N. Kinney Road
Tucson, AZ 85743-8918
Tel: (520) 883-1380
Fax: (520) 883-2500
email: fpollen@azstarnet.com

*This unique museum is situated in the Sonoran Desert of southern Arizona. Exhibits focus on the animal and plant life, geology, and history of this subtropical desert area of Arizona, California, and northern Mexico. A nonprofit, membership-supported organization active in educational outreach for students, educators, and adults, the museum conducts extensive on-site programs and trips to field sites around the world. It also publishes a newsletter,* Sonorensis, *and natural history books. It is the home of new pollinator gardens, arthropod exhibits, the Forgotten Pollinators public awareness campaign, and the Desert Alert program for volunteers.*

Bat Conservation International (BCI)

P.O. Box 162603

Austin, TX 78716-2603

Tel: (512) 327-9721

*This organization seeks to promote bat conservation by increasing public awareness of their ecological importance.*

CITES Secretariat, UNEP

15 Chemin des Anemones

Case Postale 356

1219 Chatelaine

Geneva, Switzerland

*This is a regulatory body that controls the trade in endangered species and products derived from their harvest.*

Environmental Defense Fund

257 Park Avenue South

New York, NY 10010

*This organization provides assistance with scientific, legal, and economic issues related to conservation biology and pesticide abuse.*

International Bee Research Association (IBRA)

18 North Road

Cardiff, CF1 3DY

United Kingdom

Tel: (+44) 1-222-372409

Fax: (+44) 1-222-665522

*IBRA is a nonprofit organization formed in 1949 and devoted to advancing apicultural education and science worldwide. It serves as a central clearinghouse for all published information on bees of the world, with a focus on apiculture, the study of honeybees. Its world-class library can be accessed by members and visitors. IBRA sponsors numerous international scientific colloquia and conferences and publishes scientific journals on beekeeping including* Bee World, Journal of Apicultural Research, *and* Apicultural Abstracts.

International Council for Bird Preservation
32 Cambridge Road, Girton
Cambridge, CB3 OPJ
United Kingdom
*This organization determines the conservation and protection status of bird species worldwide. Many of these birds are floral visitors and important pollinators.*

International Union for the Conservation of
Nature and Natural Resources (IUCN)
Avenue de Mont Blanc
CH-1196 Gland, Switzerland
*Also known as the World Conservation Union (WCU), this is the primary coordinating agency for international conservation efforts. IUCN publishes directories of environmental specialists who are knowledgeable about captive breeding programs and other topics.*

National Audubon Society
950 Third Avenue
New York, NY 10022
*The premier birding organization in the United States. It is also active in wildlife conservation, public educational outreach, scientific research, and other aspects of conservation.*

National Wildlife Federation
1400 Sixteenth Street N.W.
Washington, DC 20036
*A leading advocate for wildlife preservation, this organization publishes the* Conservation Directory, *as well as other outstanding educational titles including the children's publications* Ranger Rick *and* Your Big Backyard.

Natural Resources Defense Council, Inc. (NRDC)
40 West Twentieth Street
New York, NY 10011
Tel: (212) 727-2700
*NRDC uses both legal and scientific methods to monitor and influence government actions and proposed legislation affecting conservation issues and pesticide abuse.*

The Nature Conservancy
1815 North Lynn Street
Arlington, VA 22209
*This is the premier organization dedicated to the preservation of biota by means of habitat preservation through local, private, and state land acquisition and an extensive array of wildlife refuges and plant protection sites. It maintains*

*extensive databases on the distribution of rare species in the Americas, especially North America.*

North American Butterfly Association (NABA)
909 Birch Street
Baraboo, WI 53913
*This is a nonprofit organization of North American professional and amateur lepidopterists who are actively engaged in the protection and conservation of threatened and endangered lepidoptera and the habitats in which they are found. NABA educates the public about the joys of nonconsumptive, recreational "butterflying."*

Society for Conservation Biology
Department of Botany
318 Hitchcock Hall
University of Washington, Box 355320
Seattle, WA 98195-5320
*This is the leading scientific membership society for the emerging field of conservation biology. It develops and publicizes new ideas and promotes the results of field and laboratory research through its journal* Conservation Biology.

Sonoran Arthropod Studies Institute (SASI)
P.O. Box 5624
Tucson, AZ 85703-5624
Tel: (520) 883-3945
E-mail: ArthroStud@aol.com
*This nonprofit organization promotes educational outreach and research on the biology of native Sonoran Desert insects and other arthropods. SASI offers in-service training workshops for teachers, sponsors, members' events, hosts the annual "Invertebrates in Captivity" conference, publishes a newsletter,* The Instar, *and a member's magazine,* Backyard BUGwatching, *and operates the Arthropod Discovery Center in the Tucson Mountains.*

Wildlife Conservation International and New York Zoological Society
Bronx Zoo
185th Street and Southern Boulevard
Bronx, NY 10460
*These are two leaders in wildlife conservation and related research including bat research.*

World Conservation Monitoring Centre
219 Huntington Road
Cambridge, CB3 ODL
United Kingdom

*This organization monitors the global wildlife trade, the status of endangered species, the use of natural resources, and protected/conserved areas around the globe.*

World Wildlife Fund (WWF)
1250 24th Street N.W.
Washington, DC 20037

*Also known as the Worldwide Fund for Nature, this major conservation organization has branches throughout the world and is active in both scientific research and management within various national parks.*

Xerces Society
10 Ash Street S.W.
Portland, OR 97204

*This organization focuses on the conservation of insects and other invertebrates, especially butterflies. It publishes the colorful newsletter* Wings.

*Sources*

This appendix lists useful sources of pollinators, nesting materials, artificial nectar, larval host plants, and associated information. Because we do not endorse the casual introduction of pollinator species or easily managed population from one region into another, we encourage restoration ecologists and others to use this listing judiciously. A new genetic stock or nonnative species should be introduced into an area (except for totally controlled confinement situations as in some insectaries and greenhouses) only as a last resort and when no native pollinator is available—and then only after much deliberation and with all state and federal collection and quarantine permits in hand. Check your local restrictions for moving nonnative bees across state and national boundaries by first contacting your state department of agriculture and the nearest branch of the U.S. Department of Agriculture's Animal and Plant Health Inspection Service (USDA–APHIS). Otherwise you run the risk—and possible criminal penalties—of breaking state and federal regulations regarding wildlife and plant protection.

POLLINATORS

Bees West, Inc.
P.O. Box 1378
Freedom, CA 95019
Tel: (408) 728-4967
   *Commercial insectary for bumblebees* (Bombus occidentalis, B. impatiens) *and other commercial pollination services.*

Biobest/W. R. Grace
1 Town Center Road
Boca Raton, FL 33486-1010
Tel: (407) 362-1859
Fax: (407) 362-1865
*Belgian-based company selling bumblebees.*

Bio-Quip Products, Inc.
17803 LaSalle Avenue
Gardena, CA 90248
Tel: (310) 324-0620
Fax: (310) 324-7931
*Furnishes insect nets, boxes, and other supplies.*

Carolina Biological Supply Co.
2700 York Road
Burlington, NC 27215
Tel: (800) 334-5551, (919) 584-0381
*Furnishes rearing kits for painted lady larvae and various moths, as well as a wide variety of other materials, especially for Eastern U.S.*

Carolina Biological Supply Co.
Powell Laboratories Division
Box 187
Gladstone, OR 97207
Tel: (800) 547-1733, (503) 656-1641
*Same as above but for Western U.S. and Canada.*

Connecticut Valley Biological Supply Co.
82 Valley Road
P.O. Box 326
Southampton, MA 01073
Tel: (800) 628-7748
Fax: (800) 355-6813
*Offers rearing kits for painted lady larvae and various moth larvae, as well as cocoons and a variety of other materials.*

Orchard Bees
Greg Dickman
4391 County Road 35
Auburn IN 46706-9794
Tel: (219) 925-5076

*Supplier of the blue orchard bee* (Osmia lignaria), *a mason bee, along with paper straws, cardboard nesting tubes, and a videocassette on their life history and propagation. Bees are supplied during the fall, winter, and early spring as overwintering adults in their nesting tubes.*

Brian Griffin's Orchard Mason Bees
Knox Cellars
1607 Knox Avenue
Bellingham, WA 98225
Tel: (360) 733-3283

*A provider of adult blue orchard bees* (Osmia lignaria) *along with nesting materials and educational products about them. Provides starter kits called "pollinators" with three filled nests containing six to seven overwintering bees each. Also sells cardboard nesting tubes or assorted drilled pine boards as bee real estate ready to be hung on a wall and occupied by nesting females. Griffin is also the author of a popular book on this bee called* The Orchard Mason Bee.

Insect Lore Products
P.O. Box 1535
Shafter, CA 93263
Tel: (800) LIVE-BUG
Fax: (805) 746-0334

*Provides "butterfly garden" rearing kits with painted lady and buckeye larvae, as well as moth rearing kits and insect activity kits for children.*

International Pollination Systems U.S.A.
"The Pollination Company"
16645 Plum Road
Caldwell, ID 83605
Tel: (208) 454-0086
Fax: (208) 454-0092

*Provides consultant services, supplies, and breeding stock of various bees* (Megachile rotundata, Osmia lignaria, Bombus occidentalis, B. impatiens) *for commercial orchards and field crop growers.*

International Pollination Systems, Canada
Box 241
Fisher Branch, Manitoba
Canada ROC OZO
Tel: (204) 372-6920
Fax: (204) 372-6635
E-mail: 74514.3107@ compuserve.com

*International Pollination Systems' primary objective is to provide pollinating bees and assistance to growers who want to initiate their own pollination pro-*

*gram or improve an existing one. They are providers of alkali bees* (Nomia melanderi), blue orchard bees (Osmia lignaria), *bumblebees* (Bombus *spp.*), *honeybees* (Apis mellifera), *and leafcutter bees* (Megachile rotundata).

John Staples
389 Rock Beach Road
Rochester, NY 14617
Tel: (716) 865-4560

> *Offers eggs of various butterflies, including eastern black swallowtail, red ad-
> miral, orange sulphur, and common sulphur; pupae of spicebush swallowtail;
> and eggs or pupae of various moths. Also provides moth rearing kits and cocoon
> collections for schools.*

NESTING MATERIALS

D'adant & Sons, Inc.
51 South Second Street
Hamilton, IL 62341
Tel: (217) 847-3324
Fax: (217) 847-3660

> *Full honeybee supply house for beekeeping equipment. Also publishes the au-
> thoritative book* The Hive and the Honey Bee *and a periodical,* The Amer-
> ican Beekeeping Journal.

Walter T. Kelley Co.
3107 Elizabethtown Road
P.O. Box 240
Clarkson, KY 42726-0240
Tel: (502) 242-2012 , (800) 233-2899
Fax: (502) 242-4801

> *Full honeybee supply house for beekeeping materials.*

Mann Lake Supply
County Road 40 & First Street
Hackensack, MN 56452
Tel: (800) 233-6663
Fax: (218) 675-6156

> *Full honeybee supply house and distributor.*

H. J. Miller Custom Paper Tubes
P.O. Box 44187
Cleveland, OH 44144-0187
Tel: (216) 741-0378
Fax: (216) 741-3170

> *Provides fabricated-to-order paper tubes as solitary-bee nesting materials.*

Ustick Bee Board Co.
11133 Ustick Road
Boise, ID 83704
Tel: (208) 322-7778
*Supplies wooden drilled bee nesting boards in addition to modern styrofoam nesting materials for the management of alfalfa leafcutter bees* (Megachile rotundata).

Wild Birds Unlimited
1430 Broad Ripple Avenue
Indianapolis, IN 46220
Tel: (317) 251-5904
*Provides various kinds of nest boxes and seeds for native wild birds.*

NECTAR SUBSTITUTES

Biotropic U.S.A.
P.O. Box 50636
Santa Barbara, CA 93150
Tel: (805) 969-9377
*Supplies Bio-Nektar.*

Nekton U.S.A., Inc.
1917 Tyrone Boulevard.
St. Petersburg, FL 33710
Tel: (813) 381-5800
*Supplies Nektar-Plus.*

PLANT CONSERVATION ORGANIZATIONS

American Association of Botanical Gardens and Arboreta
P.O. Box 206
Swarthmore, PA 19081

Center for Plant Conservation
P.O. Box 299
St. Louis, MO 63166
Tel: (314) 577-9450
Fax: (314) 577-9465

Directory of Native Plant Societies
New England Wild Flower Society
Garden in the Woods
Hemenway Rd.
Framingham, MA 07101

Forgotten Pollinators Campaign & Desert Alert
Arizona-Sonora Desert Museum
2021 N. Kinney Road
Tucson, AZ 85743-8918
Tel: (520) 883-1380
Fax: (520) 883-2500
E-mail: fpollen@azstarnet.com

National Council of State Garden Clubs
4401 Magnolia Avenue
St. Louis, MO 63110
Tel: (314) 776-7574

National Wildflower Research Center
4801 LaCrosse Blvd.
Austin, TX 78739
Tel: (512) 292-4100

Native Seeds/SEARCH
2509 N. Campbell Avenue #325
Tucson, AZ 85719
Tel: (520) 327-9123
Fax: (520) 327-5821

APPENDIX 5

Pollinator Classes for the World's Wild Flowering Plants
(Approximately 240,000 flowering plant species worldwide)

| CATEGORIES OF POLLEN VECTORS | NUMBERS[a] BY CATEGORY | % OF TOTAL FLOWERING PLANTS (ANGIOSPERMS) POLLINATED BY TAXON |
|---|---|---|
| Wind (abiotic) | 20,000 | 8.3% |
| Water | 150 | 0.63% |
| Bees | 40,000 | 16.6% |
| Hymenoptera | 43,295 | 18.0% |
| Butterflies/Moths | 19,310 | 8.0% |
| Flies | 14,126 | 5.9% |
| Beetles | 211,935 | 88.3% |
| Thrips | 500 | 0.21% |
| Birds | 923 | 0.4% |
| Bats | 165 | 0.07% |
| All Mammals | 298 | 0.1% |
| All Vertebrates | 1,221 | 0.51% |

Note— We can estimate with fair precision the numbers of extant bee species or butterflies, etc. What is much more difficult, or impossible at the present time, is to know what fraction of the world's flowering plants are visited and pollinated by each class of biotic pollinators. Therefore, the numbers in the above table should only be taken as a starting estimate subject to further refinements in the coming years.

[a] Pollinator estimates are from Roubik (1995), Free (1993), and unpublished sources.

## Common Agricultural Pesticides

This appendix lists the common names and trade names of agricultural chemicals (insecticides) that are lethal or potentially dangerous to bees and other pollinators. Pesticides should be applied only when their use is necessary to protect crops from severe damage by insects. They should only be used with proper personal safety equipment and in strict accordance with label guidelines and laws. Labels often specify whether the material is safe for bees (honeybees) and at what season or time of day it should be applied. Remember that pesticides can drift during aerial application and pollinators may pick up sublethal doses from nontarget crops and roadside weeds. Microencapsulated formulations can be especially dangerous for pollen-collecting bees since the capsules are the same size as pollen grains. The following abbreviations are used here for pesticide formulations:

| | |
|---|---|
| D | dust |
| EC | emulsifiable concentrate |
| F | flowable |
| G | granular dry powder |
| LS | liquid suspension |
| MA | concentration applications according to mosquito abatement rates |
| SP | soluble powder |
| ULV | ultra-low-volume application |
| WP | wettable powder |

Since the following pesticides are extremely toxic to bees and other pollinators, they should not be applied to crops or weeds in bloom or when large numbers of bees are present. The residual toxicity of these formulations is often high enough to kill large numbers of bees even 12 hours or more following their application.

New formulations of pesticides appear daily, and it takes years for researchers to test them on various crops and pollinators. Thus for the latest information the reader is directed to specialty publications on bees and pesticide safety including those published by the International Bee Research Association in England.

acephate
aldicarb
aldrin
aminocarb
azinphos-ethyl
azinphos-methyl
bendiocarb
calcium arsenate
carbaryl (ULV)
carbofuran (F)
carbophenothion (D)
carbosulfan
chlorpyrifos
crotoxyphos
cypermethrin
deltamethrin
diazinon
dicapthon
dichlorvos
dicrotophos
dieldrin
dimethoate
dinoseb
DNOC
endosulfan
EPN
etrimphos
fenamiphos
fenitrothion
fensulfothion
fenthion

fenvalerate
flucythrinate
formotion
gamma-HCH
heptachlor
heptenophos
isobenzan
lead arsenate
malathion (D)
malathion (EC, ULV)
methamidophos
methidathion
methiocarb
methomyl (D)
mevinphos
monocrotophos
naled (EC, D, WP)
omethoate
oxamyl
paraquat
parathion
parathion-methyl
permethrin
phenthoate
phorate
phosmet
phosphamidon
phoxim
pirimiphos-ethyl
pirimiphos-methyl
profenophos

propoxur
quinalphos
resmethrin
sulfotep
sulprofos
tetrachorvinphos
thiometon
thionazin
triazophos
vamidothion

These pesticides should be applied to crops and other plants only during the late evening, night, or early morning to avoid coinciding with times of bee flight. Their toxicity is usually greatly reduced within 3–6 hours following application.

| | | |
|---|---|---|
| aminocarb (ULV) | dioxathion | methoxychlor |
| binapacryl | disulfoton (EC) | Nissol |
| carbaryl (ULV) | DNOC | oxamyl |
| carbophenothion | endosulfan | oxydemeton-methyl |
| chlordane | endrin | phorate (G) |
| chlorfenvinphos | ethiofencarb | phosalone |
| chlorpyrifos (MA, ULV) | ethion | pirimicarb |
| coumaphos | fenchlorphos | propoxur (MA) |
| cypermethrin | fenthion (G, MA) | pyrazophos |
| DDT | fonofos | rotenone (D) |
| demeton | formetanate | temephos |
| demeton-S-methyl | heptachlor (G) | TEPP |
| dichlorfenthion | leptophos | tetrachlorvinphos |
| dichlorvos (MA) | malathion (MA) | thiodicarb |
| dieldrin (G) | malonoben | toxaphene |
| dimetilan | menazon | trichlorfon |
| dinobuton | methomyl (LS, SP) | |

LEAST TOXIC TO BEES

The following pesticides have proven to be the least toxic of all for honeybees. (Their toxic effects on other pollinators remain largely unknown.) As such, they can be applied virtually any time of the day with reasonable safety for flying pollinators on flowering crops. Their toxicity is fairly low even immediately following application.

| | | |
|---|---|---|
| allethrin | chlordimeform | disulfoton (G) |
| amitraz | chlorfenethol | fenazaflor |
| azocyclothin | chlorfenson | fenazoflor |
| *Bacillus thuringiensis* (BT) | chlorfensulfide | fenbutatin oxide |
| toxin | chlorfentezin | fenoxycarb |
| bromopropylate | chlorpropylate | fenson |
| carbaryl (G) | cryolite | fensulfothion (G) |
| carbofuran (G) | cyhexatin | fluvalinate |
| chinomethionate | dicofol | *Heliothis* virus |
| chlorbenside | dienochlor | hydroprene |
| chlorbenzilate | diflubenzuron | isofenphos |
| chlordecone | dinocap | lime sulfur |

malathion (G)
mancozeb
mirex (G)
nicotine sulfate
permethrin

propargite
propoxur (G)
pyrethrum
rotenone (EC)
ryania

schradan
sodium fluorosilicate baits
sulfur
tetradifon
thiocyclam

## TRADE NAMES AND COMMON NAMES FOR PESTICIDES

This section lists the common and trade names for most of the agricultural pesticides in common use throughout the world (without regard to their toxicity for bees and other pollinators). Company trade names are given first while common names, where applied, may follow in parentheses.

Abate (temephos)
Acaraben (chlorobenzi-
    late)
Acaralate (chloropropy-
    late)
Acarol (bromopropylate)
Accothion (fenitrothion)
Acephate
Acricid (binapacryl)
Actellic (pirimiphos-
    methyl)
Afugan (pyrasophos)
Agrothion (fenitrothion)
aldicarb (G)
Aldrex (aldrin)
aldrin
allethrin
alphamethrin
amidithion
aminocarb (ULV)
amitraz
Anthio (formothion)
arprocarb (see propoxur)
azinphos (azinphos-
    methyl)
azinphos-ethyl
azinphos-methyl
Azodrin (monocrotophos)
Bacillus thuringiensis (BT)
    toxin
Bassanite (dinoseb)
Basudin (diazinon)

Baygon (see propoxur)
Baytex (see fenthion)
Baythion (phoxim)
Belmark (see fenvalerate)
bendiocarb
benzene hexachloride
    (gamma-HCH)
benzophosphate (phosa-
    lone)
Bidrin (dicrotophos)
binapacryl
Bromex (see naled)
bromphos-ethyl
bromopropylate
bromphos
calcium arsenate
carbaryl (D, G, ULV,
    WP)
Carbicron (dicrotophos)
carbofuran (G, F)
carbophenothion (D)
carbosulfan
Carzol (formetanate)
chinomethionate
chlorbenside
chlorbenzilate
chlordane
chlordecone
chlordimeform
chlorfenethol
chlorfenson
chlorfensulfide

chlorfentezin
chlorfenvinphos
chlorphenamidine
    (chlordimeform)
chlorpropylate
chlorpyrifos (MA, ULV)
Cidial (phenthoate)
Ciodrin (crotoxyphos)
Co-Ral (coumaphos)
coumaphos
croneton (ethiofencarb)
crotonamide (monocro-
    tophos)
crotoxyphos
cryolite
Curater
Cygon (dimethoate)
cyhexatin
Cymbush (see cyperme-
    thrin)
cypermethirin
cypermethrin
Cythion (see malathion)
Dasanit (see
    fensulfothion)
DDT
DDVP (see dichlorvos)
Decis (deltamethrin)
Delnav (dioxathion)
deltamethrin (see en-
    domethrin)
demeton

demeton-methyl
demeton-S-methyl
Derris (see rotenone)
diazinon
Dibrom
dichlorfenthion
dichlorvos (MA)
dicofol
dicrotophos
Didimac (DDT)
dieldrin, (G)
dienochlor
diethion (ethion)
diflubenzuron
Dimecron (phoshamidon)
dimethoate
dimetilan
Dimilin (diflubenzuron)
Dimite (chlorfenethol)
dinitrocresol (see DNOC)
dinobuton
dinocap
dinoseb
dioxacarb
dioxathion
Dipel *(Bacillus thuringiensis)*
Dipterex (trichlorfon)
disulfoton, (EC, G)
Di-Syston (see disulfoton)
Dithane (mancozeb, maneb, or zineb)
dithic
Dithiofos (sulfotep)
DNBP (dinoseb)
DNOC
Draza (methiocarb)
Dursban (see chlorpyrifos)
Dyfonate (fonofos)
Dylox (trichlorfon)
Ekalux (quinalphos)
Ekamet (etrimphos)
Ekatin (thiometon)
Elcar (*Heliothis* virus)
endosulfan
endrin

EPN
ethiofencarb
ethion
etrimphos
Fastac (alphamethrin)
fenamiphos
fenazaflor
fenbutatin oxide
fenchlorphos
fenitrothion
fenoxycarb
fenson
fensulfothion, (G)
fenthion (G, MA)
fenvalerate
Fernos (pirimicarb)
Ficam (bendiocarb)
flucythrinate
fluvalinate
Folbex (chlorobenzilate)
Folimat (fenitrothion)
Folithion (fenitrothion)
fonofos
formetanate
formothion
Fundal (chlordimeform)
Furadan (see carbofuran)
Galecron (chlordimeform)
gamma-BHC (gamma-HCH)
Gamma-Col (gamma-HCH)
gamma-HCH
Gammexane (gamma-HCH)
Gardona (see tetrachlorvinphos)
Gusathion (A,K) (azinphos-methyl)
Gusathion (M) (azinphos-ethyl)
Guthion (azinphos-methyl)
HCH (gamma-HCD)
*Heliothis* virus
heptachlor (G)

heptenophos
hydroprene
Hostathion (triazophos)
Imidan (phomet)
isobenzan
isofenphos
Karathane (dinocap)
Kelthane (dicofol)
Kepone (chlordecone)
Kilval (vamidothion)
Lannate (see methomyl)
Larvin (thiodicarb)
lead arsenate
Lebaycid (see fenithion)
leptophos
lime sulfur
lindane (gamma-HCH)
Lorsban (see chlorpyrifos)
malathion (D, EC, G, MA)
mancozeb
Matacil (aminocarb)
Maverick (fluvalinate)
mecarbam
menazon
mercaptothion (see malathion)
Mesurol (methiocarb)
Metasystox (demeton-S-methyl)
Metasystox-R (oxydemeton-methyl)
metathion (fenitrothion)
methamidophos
methidathion
methiocarb
methomyl (D, LS, SP)
methoxychlor
methyl-demeton (demeton-S-methyl)
mevinphos
Milbex (chlorfensulfide with chlorfenethol)
mirex (G)
Mitac (amitraz)
Mitox (chlorbenside)

Monitor (methamidophos)
monocrotophos
Morestan (chinomethionate)
Morocide (binapacryl)
Murfotox (mecarbam)
naled (D, EC, WP)
Neguvon (trichlorfon)
Nemacide (dichlofenthion)
Nemacur, P (fenamiphos)
Nemaphos (thionazin)
nicotine sulfate
Nogos (see dichlorvos)
Nuvacron (monocrotophos)
Nuvan (see dichlorvos)
Oftanol (isofenphos)
omethoate
Omite (propargite)
OMPA (schradan)
Orthene (acephate)
Ovex, Ovotran (chlorfenson)
oxamyl
oxydemeton-methyl
oxythiquinox (chinomethionate)
parathion
parathion-methyl
Pentac (dienochlor)
perchlordecone
permethrin
phenthoate
phorate (EC, G)
phosalone
Phosdrin (mevinphos)
phosmet
phosphamidon
Phosvel (leptophos)
phoxim

pirimicarb
Pirimicid (pirimiphos-ethyl)
pirimiphos-ethyl
pirimiphos-methyl
Pirimor (pirimicarb)
Plictran (cyhexatin)
Pounce (see permethrin)
profenophos
propargite
propoxur (G, MA)
prothiophos
Pydrin (see fenvalerate)
pyrazophos
pyrethrum
quinalphos
quinomethionate (chinomethionate)
resmethrin
Ripcord (see cypermethrin)
Rogor (dimethoate)
ronnel (fenchlorphos)
rotenone (D, EC)
ryanla
Ryanodine (ryania)
Sapecron (chlorfenvinphos)
Sayfos (manazon)
schradan
Sevin (see carbaryl)
sodium fluorosilicate
Solvigran, Solvirex (see disulfoton)
Spur (fluvalinate)
stirofos (see tetrachlorvinphos)
Strobane (toxaphene)
sulfotep
sulfur
sulprofos

Sumithion (fenitrothion)
Supona (chlorfenvinphos)
Supracide (methidathion)
Systox (demeton)
Sytam (schradan)
Tamaron (methamidophos)
tecnazene
Tedion (tetradifon)
Telodrin (isobenzan)
temephos
Temik (G) (aldicarb)
TEPP
Terracur (see fensulfothion)
tetrachlorvinphos
tetradifon
Thimet (see phorate)
Thiocron (aminithion)
thiocyclam
Thiodan (see endosulfan)
thiodicarb
thiometon
thionazin
thiotep (sulfotep)
toxaphene
triazophos
trichlorfon
Trithion (see carbophenothion)
Ultracide (methidathion)
Unden (see propoxur)
Valexon (phoxim)
vamidothion
Vapona (see dichlorvos)
Velsicol (see heptachlor)
Vendex (fenbutatin oxide)
Vydate (oxamyl)
Yaltox (see carbofuran)
Zinophos (thionazin)
Zolone (phosalone)

* Neither the authors, the staff of the Arizona–Sonora Desert Museum, nor any members of the Forgotten Pollinators campaign endorse any of the aforementioned insecticides or formulations. Any agrichemicals should only be applied when necessary and only applied in a safe manner according to the label restrictions and directions.